About the Authors

Dr Simon Torok works in communication and marketing for CSIRO. He has a PhD in Earth Sciences from the University of Melbourne and a Graduate Diploma in Science Communication from the Australian National University (ANU). Simon has worked as editor of CSIRO's Helix magazine, performed with the ANU/Questacon Science Circus, and as a climate communicator in the United Kingdom. He has done science talkback segments on radio and has published dozens of newspaper, magazine and scientific journal articles. He hopes one day to invent a car radio that features a 'record' button, and would really like to see someone develop an analogue clock radio.

Paul Holper works for CSIRO in business development and communication. He has an Honours Degree in Chemistry, a Diploma of Education, and a Graduate Diploma in Science Communication. Before joining CSIRO, Paul taught secondary school chemistry and science. He often appears in the media describing Australia's latest scientific advances, and has written numerous newspaper and magazine articles as well as five science textbooks and a CSIRO Division history. The invention that Paul craves is a pill that instantly gives you full command of subject areas or skills. He would immediately take a French-speaking capsule and a tennis-playing tablet.

Paul and Simon are also the authors of the *Amazing Science* series, published by ABC Books: *Wow! Amazing Science Facts & Trivia*; *Whiz! Amazing Maths & Science Puzzles*; *Zap! Amazing Science Experiments*; and *Weird! Amazing Inventions & Wacky Science*; as well as *101 Great Australian Inventions*, *101 Great Solar System Facts & Trivia*, and *101 Great Killer Creatures*.

Inventing Millions

25 inventions that changed the world (and made millions for their inventors)

Simon Torok & Paul Holper

From the authors of the bestselling Amazing Science series

illustrations by Danny Snell

ABC Books

Published by ABC Books for the
AUSTRALIAN BROADCASTING CORPORATION
GPO Box 9994 Sydney NSW 2001

First published in June 2006

National Library of Australia
Cataloguing-in-Publication entry
Torok, Simon.
Inventing millions: 25 inventions that changed the world
(and made millions for their inventors).

1st ed.
Bibliography.
Includes index.

ISBN 10: 07333 1714 6.
ISBN 13: 978 07333 1714 9.

1. Inventions. 2. Technological innovations.
I. Holper, Paul N., 1957–. II. Title.

608

Cover and internal design by Luke Causby, Blue Cork Design
Typeset in 10 on 13pt Sabon by Agave Creative Group
Printed and bound in Australia by Griffin Press

5 4 3 2 1

To Matilda Torok – ST
In memory of Alex Holper – PH

Acknowledgements

Thanks a million to Helen Doyle, Janet Holper, Andi Horvath, Tom Montague, Margaret Snare and David Salt for their valuable comments. We are also indebted to Helen Littleton and Brigitta Doyle at ABC Books, and to Sandra Goldbloom Zurbo for her superb editing.

ABC Books acknowledges Lothian Books for permission to quote from *MacRobertson: The Chocolate King* by Jill Robertson.

Contents

Foreword

We humans are one of the few animals that use tools. In fact, if you get right down to it, the two basic tools are a hard rock and a soft rock – in other words, the hammer and the file. With the hammer, you can make something change shape, and if you hit it hard enough and long enough you can even weld metals together (it helps to have a fire). With a file, you can remove any bits of metal you don't want. So an engineer could rebuild society after a terrible disaster armed with nothing more than their brain and a few rocks – and after a few years would have everything from medical devices to jet engines again.

Many inventors came up with something useful to use around the house as their very first invention. From this, not only did they get fabulously wealthy, they helped make our world a better place. I hope that some of the readers of this magnificent book will follow in their footsteps.

Dr Karl Kruszelnicki, AM
University of Sydney's Julius Sumner Miller Fellow

Introduction

Picture a scene in your not-too-distant future. You are awoken by a tune downloaded to your mobile phone, which you hear clearly, thanks to your bionic ear, behind which you dab some Chanel No. 5. You take a swig of Red Bull to wake up and heat your breakfast, which you have stored in some Tupperware, in the microwave. Then you chew gum that restores decayed areas on your teeth (too much chocolate?). Maybe you do a bit of vacuuming with your Dyson and cleaning with ENJO microfibre gloves before riding your superbike, purchased on your Diner's Club credit card, past a stream of cars to work. Once you arrive at your office, you do a Google search on your nanotechnology computer for the cost of a holiday in space, noting the details with a space pen. You make it through the day, thanks to your pacemaker and, when you arrive home, open a can of widget-fizzed Guinness to toast your many accomplishments.

Technology has transformed our world and our way of life – although not always to the extent of the scenario above, which features twenty of the twenty-five inventions described in this book. *Inventing Millions* tells the stories of how innovation, discovery, science and technology have transformed millions of people's lives – medically, by saving millions of lives with the discovery of penicillin and the invention of the pacemaker, financially, by turning the inventor into a millionaire, as have items such as the Rubik's Cube and Tupperware, by making millions for an inventor's company, as have the microwave oven and the mobile phone, by reaching millions of people, as the digital music revolution and Google have or, potentially, changing millions of lives, as could nanotechnology.

Inventions have often handsomely rewarded the scientists, engineers and entrepreneurs who had the wisdom, drive and luck to turn ideas into reality. Scanning through Top 100 and 100 Richest lists the reader comes upon names across a range of industries – IT, real estate, entertainment, technology – all of which require some degree of innovation and have been responsible for many million dollar achievements.

The question is, what does it take for you to be the next person to make or save millions? In researching and writing this book, we had hoped to find links and common themes underpinning some of the most successful product developments. Is there a formula for success that might help budding inventors and entrepreneurs make their millions? The answer is yes, sort of. We discovered that the emergence of many of the twenty-five creations we examined had distinct similarities. The initial spark of an idea or the solution to a problem can hit while on holiday (as it did with the bionic ear), in a dream (where it happened for the space pen), through a chance observation (as with penicillin and microwave ovens), by sheer hard work and determination to solve a problem (such as occurred with the non-clogging vacuum cleaner), or through investment in research and development to find your idea, which was how Guinness developed its widget.

You must believe in your product – the inventor of a bulletproof vest allowed himself to be shot at. Persistence pays, as inventions are usually rejected by the first companies approached, as was the case for Rubik's Cube and the Dyson vacuum cleaner. Good marketing, too, is essential –in the case of Red Bull, it was arguably the most important factor – and may involve hitting the road to demonstrate your product, which is how it was done for Liquid Paper, microwave ovens, Rubik's Cube and Tupperware. Just ensure, though, that, once you are successful, you avoid the temptation to run for government – successful inventors Ford (cars), Fisher (pens) and MacRobertson (chocolate) all failed in their bids for political office.

Hungarian Nobel Prize winning scientist Szent-Gyorgyi believed that discovering is seeing what everybody sees, but thinking what nobody else thinks. This could simply mean recognising a good idea and improving it. Henry Ford took the already-invented motor car and the production line and improved them both to make his fortune. With a high failure rate for new products and an even higher failure rate for new ideas, sometimes it is better to be second, not first, and to build on a proven idea.

Not all inventions improve on a previous idea or meet an obvious need. Some things come out of nowhere to become an unexpected hit. Who would have thought that so many people would want a Rubik's Cube? Albert Einstein said imagination is more important than knowledge – knowledge is limited, but imagination spans the world.

Great things can be achieved by each of us if we think, persevere and follow our dreams. In future perhaps the big ideas will come from small companies or individuals. The internet is connecting people around the world in ways that were unimaginable a decade or two ago. Just as the digital music revolution is enabling anyone to distribute their original music from home, the internet could enable an ideas revolution, leading to virtual teamwork on innovations from around the world. As Scottish scientist Alexander Fleming pointed out, the lone worker comes up with the prime idea and makes the first advance in a subject; the details are worked out by a team.

Of course, if reading books on making millions doesn't work for you, you could always try writing one with the aim of selling millions. You wouldn't be the first.

Good luck becoming one-in-a-million.

Simon Torok and Paul Holper

Inventing Millions

Bigger is Better?
No, Smaller is Superior

How wireless communication and the mobile phone made the world smaller

Nineteen seventy-three – a year of many advances in technology. The *Viking* spacecraft lands on Mars, the Sydney Opera House opens after years of construction and the oil crisis precipitates plans for innovation in energy conservation. Meanwhile, Bob Marley is popularising reggae music and Richard Nixon is president of the USA, but won't be for much longer as two of his aides have been convicted over the Watergate break-in.

On 3 April of this same year, a Chicago-raised man with an electrical engineering degree from the Illinois Institute of Technology steps onto a New York street. He then does something that has arguably influenced our lives to a greater extent than any of the milestones mentioned above.

The man's name is Martin Cooper, and he is 43 years old. Martin later explains to us by email that he always had an insatiable curiosity about how things work.

'The navy put me through school, after which I ultimately became a submarine officer,' he says. 'That experience brought some reality into my life.'

Perhaps it is his time on the submarine that makes him appreciate compact equipment and efficient use of space. Perhaps his confinement in a submarine has also given Martin an appreciation of conducting his activities in open spaces.

Whatever the reason, Martin Cooper is now standing outside on a street near the Manhattan Hilton. He has travelled to New York to demonstrate his invention to the public, having first tested it in Washington DC. The invention had been developed by the research team he leads at a large telecommunications company, a position he rose to after building

products such as portable police radios. The company is Motorola. In Martin's hand is the world's first personal, handheld cell phone – and he is about to use it for the first time.

Who you gonna call?

Of course, Martin's mobile phone was quite different to the ones used now. Mobile phones today list hundreds of functions from camera to karaoke. In 1973, Martin's phone had just three functions – you could dial, you could talk, and you could listen.

Whose number should he dial on this momentous occasion? Nobody else has a mobile phone yet, so that narrows it down. He needs to call someone whom he knows will be at home or in an office. How about Joel Engel, head of research at Bell Laboratories and Martin's main rival?

Martin picks up the handset and presses the 'off-hook' button, bringing the phone to life. A signal is sent from the phone to a base station, which the Motorola team have installed on the roof of a nearby building. The signal then connects to the landline system. Back on the street, Martin dials the number and puts the handset to his ear.

'Hi, Joel. Guess where I'm calling from.'

In this historic conversation, Martin explained to Joel that he was on the street, speaking from a hand-held cellular phone. Today, he doesn't remember what Joel replied, but he jokes that he thinks he heard the gnashing of teeth.

Wireless revolution

The conversation signalled success for Martin's vision for wireless communication for people, as opposed to the mobile calls that could already be made from cars or trains.

'I learnt at Motorola that the freedom inherent in truly personal communications was something special,' Martin reflects in an email to us.

It could not have been too long a conversation as the battery lasted only 10 to 20 minutes. Just as well, as the phone could be held up for only 10 minutes because of its weight. It was huge. It looked like a large car battery with a handle and phone handset. Weighing more than a kilogram, you could develop muscles just by holding it.

As he spoke into this large contraption, passers-by looked on in bewilderment. Martin describes the experience on the website of his current company, ArrayComm. 'As I walked down the street while talking on the phone, sophisticated New Yorkers gaped at the sight of someone actually moving around while making a phone call. Remember that in 1973, there weren't cordless telephones, let alone cellular phones.' Preempting the dangers of speaking on a mobile phone while driving, Martin says, 'I made numerous calls, including one where I crossed the street while talking to a New York radio reporter – probably one of the more dangerous things I have ever done in my life.'

First phones

Back in 1973, Martin could never have dreamed of more than a billion people around the world owning a mobile phone. Fortunately, Motorola showed more foresight than a Western Union company executive had done 100 years earlier with a related invention – the world's first telephone. An internal memo declared that 'this 'telephone' has too many shortcomings to be seriously considered as a means of

Early days:
Martin, Morse and more

Even though Martin is widely recognised as the father of the cell phone, Isaac Newton's quote comes to mind: 'If I have seen further it is by standing on the shoulders of giants.' This suggests two things. First, that Newton may not have been a very tall man. Second, that it is worth a quick discussion of mobile phone pre-history to fully understand Martin's place in it.

Samuel Morse invented the telegraph in 1837 and soon after, we assume, the code required to use it that bears his name. Italian inventor Antonio Meucci developed early telephones in the mid 1800s in New York but Alexander Graham Bell, who shared a lab with Antonio, acquired the patent and international fame after his 1876 landline call and subsequent radio telephone voice call in 1880. In 1901, Marconi sent radio signals from England to America across the Atlantic. A simple car telephone was invented in 1910 by Lars Magnus Ericsson, who had previously founded the Ericsson company in 1876. In the early 1920s, when Morse code was still the main form of radio communication, the Detroit Police Department pioneered the use of a one-way radio paging system for police cars. But it wasn't until the mid 1950s that a phone-equipped civilian car took to the road, and even then it was like carrying a complete telephone exchange around in the boot. The first commercial cellular phone calls were made in 1969 on AT&T Bell's telephones aboard Metroliner trains running between New York and Washington DC.

In 1947, AT&T Bell had proposed to the government body regulating the American airwaves that a number of radio frequencies be allocated to a widespread mobile phone

Early days cont.

network. But enough space was granted to allow only twenty-three simultaneous phone conversations, so there wasn't much incentive for research. In the 1960s, AT&T proposed a cellular system and the government regulator increased the network's phone-carrying capacity. The race was then on to develop the concept into a technologically and economically feasible product.

These advances set the scene for Martin Cooper's New York curbside call in 1973.

It is worth finishing this history with a description of an uncannily similar call made nearly 100 years earlier. In 1879, David Hughes walked along Great Portland Street in London with a telephone to his ear, listening for the electrical clicks sent by a transmitter in his house. But David did not pursue the invention because, following a demonstration in 1880 to the Royal Society, his unimpressed colleagues claimed the signals were not carried by radiowaves. As described in an account of the streetside experiments published twenty years later, 'We are not told what passers-by thought of the learned scientist, apparently wandering aimlessly about with a telephone receiver held to his ear, but doubtless they had their own ideas.' Although not voice transmissions, the clicks *were* radiowaves, the precursor to mobile phone carrier waves. This experiment can hence be considered the first ever mobile phone call.

communication. The device is inherently of no value to us.'

In 1983, ten years after Martin's first call, Motorola introduced onto the market the Dyna-Tac. Although half the size of Martin's prototype, it still looked like a large brick with a keypad. 'The first phones built in quantity, even then, cost $4000,' Martin recalls – that's the equivalent today of more than $10 000 a phone. Another ten years later, mobile phones would be much cheaper – often even free when given away as part of a package deal.

Today, there are more people around the world with cellular phones than there are with landline phones. It's not at all surprising to see two people sitting in a cafe talking not to each other but into their phones, or to watch others walking along the street engrossed in writing a text message. With more than 150 million new people signing up for a mobile phone each year, the total number of mobile phone subscribers around the world will soon reach 2 billion. In Australia alone, with a population of just over 20 million, there are more

---

You're kidding:
Cellular culture

Before cellular phones appeared in the real world, they featured in pop culture from cartoons to sitcoms. In the 1960s television show *Get Smart*, Secret Agent 86 used a phone built into his shoe. Even more advanced than Maxwell Smart's shoe-phone was Dick Tracy's voice-activated, wrist-based videophone, which appeared in the 1940s eponymous cartoon. Max and Dick could make calls from anywhere because they always had the devices with them. Being fixed to the Batmobile, the portability of the Batphone used by Batman and Robin pales in comparison.

Once fact caught up with fiction and mobile phones appeared in movies set in the present day, they became a means of dating a movie. You can usually tell the year in which the film was made by noting the size or even the model of the mobile phone used by the movie's stars.

HOW IT WORKS:
Back to base

A mobile phone converts vibrations from your voice to electrical signals, just as a landline does, but the mobile phone converts these signals to radiowaves, unlike a landline phone, which sends the electrical signals along a telephone line. This is analogous to the way a picture is sent from a television station, where the electrical signal is mixed with a carrier wave produced by an electromagnet. The carrier waves are transmitted in all directions from the antennae, just as waves ripple in a pond, but the difference being that carrier waves travel at the speed of light. The signals are received at a nearby base station or mobile phone tower.

The mobile phones most of us use are called cellular because they are part of a system that uses numerous base stations to divide the service area into many small regions called cells. As you travel from cell to cell, the calls are transferred from base station to base station, which sends the signal to other base stations and eventually to a central telephone exchange, which distributes the message to the number being called.

than 17 million mobile phone users, generating almost $9 billion a year in mobile phone service revenue.

While Motorola has made millions from the technology, Martin didn't get rich from his invention.

'When I joined Motorola I sold my interest in any inventions for $1, which I still have,' he says.

He didn't even receive a bonus for building the first truly mobile phone. It was just a part of his job.

So we take this as a lesson in how not to be a millionaire. But he does question the definition of 'rich'. 'The satisfaction of achievement was more valuable than any cash,' he says. Martin has used his fame to become a successful entrepreneur in other endeavours. He heads a Californian wireless technology firm, ArrayComm, which has created the fundamental technology to expand personal communication to computers, cameras, games and medical devices.

SUCCESS SCALE

INNOVATION

Martin's invention made an old idea better by taking the existing technology of the phone and making it mobile. Martin says his advice to other entrepreneurs comes from his mentor, Motorola founder Paul Galvin: 'Reach out. Do not fear failure.'

MARKETING

The mobile phone met a perceived need, so marketing was directed to people who wanted to be mobile and felt they needed a mobile phone as soon as they saw one.

FINANCE

The mobile phone has made millions for Motorola and other phone companies, as well as many of the developers in related industries. However, because Martin was an employee, he had to rely on other ideas after he left Motorola to make a million himself.

References

Internet

<www.arraycomm.com>.
<inventors.about.com>.
<www.privateline.com>.
Encyclopaedia Britannica online, at <www.britannica.com>.

Scent of the Century

How the perfume tree helped Coco Chanel climb to the top of the fashion industry

Saumur is a peaceful, pretty town in France's Loire Valley. It is famous for its spectacular scenery and its fine sparkling wines. Dominating the town is the splendid chateau built by a tenth century count and renovated by nobility innumerable times in the centuries that followed.

On 19 August 1883, almost a millennium after construction of the Chateau de Saumur, a baby girl was born in the town's hospital for the poor. Today, few people know Saumur, but there are few who would not know of the empire that the little girl would grow up to create. Gabrielle Bonheur Chanel lost her mother when she was young. Her father, left with five children to raise, quickly farmed them out to relatives before disappearing. Details of Chanel's childhood are scant, but we know that, as well as living with relatives, she also spent time in a local orphanage. Nicknamed 'Coco', or 'Little Pet', it was her aunts who taught her to sew, a

Early days:
The emperor scent me

The ancient Chinese, Hindus, Israelites, Arabs, Greeks and Romans all created and used perfumes. Pyramid building was sweaty work. Egyptian papyrus manuscripts from around 2700BC describe the use of fragrant herbs, choice oils, perfumes and temple incense to sweeten bodies and the air.

The bible says, 'Ointment and perfume rejoice the heart' (Proverbs 27: 9). You'll also find recipes for perfumes in the bible.

First-century Rome was importing thousands of tonnes of frankincense and hundreds of tonnes of myrrh a year. Frankincense, a resin from select trees, is rich in volatile oil that produces a strong fragrance as it burns. Myrrh is a mixture of resin, gum and essential oil that was highly valued as an ingredient of perfume and incense.

Emperor Nero might have fiddled while Rome burnt, but he certainly knew how to party. Nero had the local plumbers fit concealed pipes into the ivory ceilings of his dining rooms. Once his parties were jumping, mists of fragrant waters engulfed his guests. Simultaneously, masses of sweet-smelling roses descended. When they weren't partying, or perhaps in preparation for a night at Nero's, Romans anointed their bodies with unguents, scented oils and powders.

skill that would help set her on the path to fame and fortune.

The nuns at the orphanage helped the teenage Coco obtain work as a seamstress. She was 18 when she left to work in a hosiery shop in Moulins in central France. But Coco had far grander plans. Singing, not sewing, is what appealed to Coco. She sought a career as a cabaret singer, soon discovering the benefits that could be had from befriending the high-society folk who frequented nightclubs. A relationship with

Etienne Balsan, a well-liked sportsman and horse breeder, led to him financing Coco's move to Paris. There, from Etienne's apartment, she sold simple, decorated hats. With financial help, Coco was soon able to open a small hat shop. By 1913, there was a second shop, this one in the resort town of Deauville.

Coco was quick to expand from hats to sweaters, skirts and accessories. She had been making her own clothes from unconventional materials, such as jersey, a clingy silky material. At the time jersey was used primarily for workers' clothes and men's underwear. Coco's distinctive 'poor girl' look proved popular with wealthy Parisians. Here was a stunning departure from prevailing frills, flounces, corsets and long skirts. 'Poor girl' transformed Coco to rich girl; her

nonconformist, simple, comfortable designs set trends for decades.

Coco's clothes also reflected the sombre times associated with the ravages of World War I. During the period, she moved into a larger shop on rue Cambon, near the Ritz Hotel where she sold flannel blazers, linen skirts, jersey sweaters and simple tailored jackets.

No. 5

In 1921, with business booming, Coco commissioned a perfume to add to her rapidly expanding product range. Its creator was a Russian perfumer, Ernest Beaux, whose creation, Chanel No. 5, would ensure that Coco Chanel's name would live on. Ernest elected to introduce a key ingredient into his new scent, a synthetic compound produced by the chemical industry – floral aldehydes. This was the top note, the initial, refreshing scent.

Ylang Ylang, the oil that results from steam-distilling flowers from the South Asian perfume tree, is another top note ingredient. At the risk of detracting from the mystique of what is, arguably, the world's most famous scent, the perfume tree is a member of the custard apple

family. The tall, slender evergreen is adorned year-round with drooping, rich-scented, greenish-yellow flowers.

The perfume professors say that Chanel No. 5's middle notes – the character builders – include jasmine, iris and lily of the valley. The persistent base note comprises sandalwood, oak, musk and chive. Somewhere mixed in among all these notes are rose, citrus and vanilla.

Why No. 5?

The story, almost certainly apocryphal, is that Ernest mixed multiple perfume samples for Coco's assessment, labelling them by number. Coco selected bottle number 5. More likely is the account that five was Coco Chanel's lucky number. And what an extraordinarily lucky number it turned out to be. Coco was the first designer to lend her name to a perfume. If you couldn't afford the Chanel clothes, you could at least save up for some No. 5.

A simple but elegant little square Art Deco design bottle completed the new perfume package. The bottle shape endures to this day.

Launched in 1922, the liquid gold helped support Coco's future business activities. At its peak, Chanel industries employed 3500 people and included a fashion house, textile business, perfume laboratories and a costume jewellery factory.

In 1924, Coco entrusted the manufacture and distribution of her perfumes to businessman Pierre Wertheimer; she retained 10 per cent of the business. Pierre's descendents still run the company.

Throughout the 1920s, Coco launched a variety of other perfumes. There was Chanel No. 22, also released in 1922 (no prizes for coming up with a rationale for that Chanel number), Cuir de Russie, Gardenia and Bois des Iles.

Marilyn Monroe gave Coco's perfume a huge boost 30 years after its launch. In 1953, when asked what she wore in bed, Marilyn replied, 'Why, Chanel No. 5 of course!'

To commemorate Coco's eighty-seventh birthday – 19 August 1970 – the company released Chanel No. 19. Coco died a few months later in her apartment at the Ritz Hotel, across the road from where her business journey had begun some 50 years earlier in a hat shop in rue Cambon.

HOW IT WORKS:
Your nose knows

Perfumes can contain natural and synthetic materials. Heat the petals or leaves of some plants and you obtain their essential oils. Like any branch of science, or witchcraft for that matter, perfumery comes with its own unique processes and terms. *Enfleurage* involves placing petals in animal fat. The fat absorbs the flower oil. Alcohol dissolves the good oil from the fat. Use hot fat and the process is called *maceration*. Distil essential oils and you can collect individual components at their unique boiling temperatures.

The perfume industry has a sad history of exploiting animals as feedstock. Some animal secretions fix perfumes, making them persist longer on your skin. The products include ambergris from sperm whales, castor from beavers, civet from civet cats and musk from deer.

Individual perfumes may be mixtures of more than a hundred components, some present in minute quantities. The dominant odour can be floral, spicy, woody, mossy or herbal. Citrus, spice and lavender are often key scents in men's fragrances.

The components are dissolved in alcohol. Expect up to 25 per cent concentrates in perfume, up to 6 per cent in eau de toilette or cologne and up to 2 per cent in aftershave.

Most people can distinguish between 3000 and 10 000 different smells, some at vanishingly small concentrations. Diphenyl sulfide, for example, has an unpleasant smell that your nose will detect at 48 parts per trillion. This means that just a drop or two of diphenyl sulfide vaporised and released into, say, a school basketball stadium would soon have almost everyone crying foul.

You're kidding:
Pheromones moans

Scents may affect you in ways you can't control. Pheromones are chemicals released in tiny quantities by animals, insects and even algae and fungi. Ants mark trails with pheromones. Injured minnows release them to warn their minnow mates that danger may be lurking. Female silk moths release pheromones that will attract males from over 2 kilometres away.

It's controversial, but humans may also be susceptible to the lure of the pheromone. We mammals detect pheromones via a tiny vomeronasal organ hidden inside our noses. We can't smell pheromones, but our little organs can detect them. And when they do, they may tell your body to release sex hormones known as gonadotropins. Researchers don't know whether the released gonadotropins affect your mood, behaviour or sex appeal, but perfume makers aren't taking any chances. As well as including scents that are supposed to inflame your desire, some perfumes contain human pheromones, claiming that they'll drive women or men wild with desire.

Scientists have found fatty acids in vaginal secretions that are identical to those thought to be sex pheromones in other primates. Women's sensitivity to musky odours is greatest around ovulation and is perhaps associated with the musky pheromones released by men. Pheromones may well be responsible for synchronising the menstrual cycles of women living or working together. But the jury is still out on the link between pheromones and human sexual response.

The Archives of Sexual Behaviour reports a 1998 study led by Winifred Cutler. The research team signed up thirty-eight heterosexual men and examined the effect of exposing them to synthesised human male pheromones. The exposed men were more likely to engage in petting, kissing and sexual intercourse with a woman than were those receiving a placebo.

You're kidding cont.

A 2003 University of California study described in *Behavioral Neuroscience* explored the influence of a natural sex steroid in human male sweat. Twelve heterosexual male and twelve heterosexual female students joined in. High concentrations of the male steroid induced small but significant increases in arousal in women.

SUCCESS SCALE

MARKETING

Link the product to a gorgeous movie star. What else can they do?

THE FUTURE

Chanel No. 5 is entirely reliant on the effectiveness of the company's marketing. Their success depends on convincing women to wear the same product as their mums.

References

Internet

Toronto Fashion Monitor online, at <www.toronto.fashion-monitor.com/designers.php/chanel>.

Time online, '100 Artists and Entertainers', at <www.time.com/time/time100/artists/profile/chanel2.html>.

Wikipedia <en.wikipedia.org/wiki/Chanel_No._5>.

Encyclopaedia Britannica online, 2001, at <www.britannica.com>.

Texts

Bensafi, M. et al. 2003, 'Sex-steroid derived compounds induce sex-specific effects on autonomic nervous system function in humans', *Behavioral Neuroscience*, 117 (6), 1125–34.

Keeping Pace

Wilson Greatbatch had a heart — and gave one to millions of others

It was 10.30 on a bitterly cold Tuesday night. Paul had just arrived home from an enjoyable night of playing bad tennis in a local competition. He was busily describing the ill-fated fightback in the second set when the phone rang.

There are phone calls you fear, phone calls you dread, phone calls that, in an instant, turn your life upside down. One minute your mind is quietly going about its business, ambling its way through a tenuously-linked assortment of recollections, aspirations, hopes and fears, while slowly erasing all knowledge of where you placed the car keys. The next minute you can think of only one thing.

Paul's mother, remarkably composed, broke the news that Alex, Paul's father, had collapsed, and was unconscious on the bedroom floor. She had just called the ambulance.

Paul leapt into his car. Fifteen minutes later, he arrived at his parents' house to find the ambulance parked outside, with a second, MICA ambulance just pulling up, siren screaming. This was grim. Racing inside, Paul found the ambulance officers, laden with equipment, desperately working to resuscitate his dad. No response, no response, no response. One of the officers, a big, burly man, grabbed defibrillator paddles and positioned them on Alex's chest. Activation. Electrical signals surged through Alex's body. Immediately, there was a tremor, a small movement. A fraction of the tension in the room lifted. This had to be a positive sign. Alex's body began to stir ever so slightly. The ambulance officers were pretty pleased. They had every right to be. 'It's not often that we can revive someone in this situation,' the burly one later admitted to Paul.

It's one thing to get a heart beating again, a different matter entirely as to whether it would keep pumping of its own volition. Then there was the question of what damage might have been sustained for however long that pump had ceased. The paramedics stretchered Alex into the back of one of the ambulances. With lights flashing, they roared off to the hospital.

Thankfully, Alex made a rapid and perfect recovery. The rapid part was fortuitous; the perfect was due to a wonderful life-saving invention, the cardiac pacemaker.

Although Alex's heart was beating, the doctors needed to ensure that there wasn't going to be another disruption to the sinoatrial node, which initiates heartbeats. So, a few days later, under local anaesthetic, a surgeon made a slit about 5 centimetres long in the skin of Alex's chest near his collarbone. Using fluoroscopy X-ray to guide the manoeuvre, the surgeon poked a fine wire through a thick vein until it reached the heart. Once this was done, he connected the other end to the pacemaker – a thin, round device sealed in a titanium case about 6 centimetres in diameter and weighing 30 grams. The surgeon carefully placed the pacemaker inside Alex's chest. Finally, he stitched the skin together and inspected his handiwork. The procedure took little more than an hour.

Immediately, the pacemaker began regulating Alex's heartbeat. When necessary, it generated a small electrical signal known as a pacing pulse that passes along the wire to the heart initiating a beat. Powered by a lithium iodide battery, this marvellous little instrument quietly and continuously senses and records heart-

Early days:
Stimulating research

In 1949, researchers at Toronto's Banting and Best Institute were studying the impact of extreme cold on heart rate, while idly wondering about the origins of their laboratory's distinctive name. (Was Banting better than Best?)

Frigid conditions could slow the human heart rate sufficiently to enable doctors to perform open heart surgery. The trouble was, sometimes hearts stopped completely. John Hopps, an electrical engineer, while trying to devise a way of rapidly restoring body temperature, realised that a small electrical stimulation of the heart would often get it beating again. What's more, by controlling the strength of the electrical signal, he could increase or decrease the heart rate.

Here was the world's first pacemaker. It was large – about 30 centimetres long, and several centimetres high and wide. Vacuum tubes generated the life-preserving pulses. If you are wondering how – or perhaps even where – surgeons managed to implant a device like this, rest assured that this was no internal device. Power came from an electrical socket. John Hopps's external pacemaker helped patients recovering from open heart surgery until their heart's own pacemaker kicked in.

Some years later, researchers developed a wearable, external pacemaker powered by batteries.

In a sense, John Hopps saved himself. Thirty years after his invention, when John's heartbeat became erratic, an implanted pacemaker solved the problem.

There are some reports that it was really an Australian doctor who should have been given the credit for inventing the first heart pacemaker. He used it in 1926 to revive a new-born baby at Crown Street Women's Hospital in Sydney. But he wanted to remain anonymous, so nobody knows his name. Thus, modesty means that he misses out on top billing in this story.

beat and, when necessary, fires off electrical prompts to the heart.

The implantable pacemaker

In the late 1940s, John Hopps, a Canadian electrical engineer, made the very first pacemaker after realising that small electrical stimulations could induce a still heart to beat again. Hopps' pacemaker was used externally as it was far too bulky to be inserted into a person's body.

By 1958, miniaturisation and development of transistors and reliable nickel-cadmium batteries meant that surgeons were able to provide a 43-year-old Swedish man named Arne Larsson with the world's first implanted pacemaker. Arne suffered from life-threatening seizures. Internal pacemaker number one lasted for just three hours. Thankfully, Arne lasted far longer, living a long and active life until, twenty-six pacemakers later, he died in 2000 at the age of eighty-six.

Who did Alex and Arne have to thank for their implantable pacemakers? Wilson Greatbatch.

Wilson Greatbatch – Bill to his friends – was born in 1919 in Buffalo, New York. After serving with the navy during the Second World War, he worked for a time as a telephone repairman, and later qualified as an electrical engineer.

Wilson was fascinated by medical science. Cornell University's psychology department employed him as an electronics technician in their animal behaviour section, where he monitored the blood pressure, heart rate and brainwaves of sheep and goats.

At Cornell, Wilson met two visiting surgeons who were performing experimental brain surgery on animals. He later explained that the surgeons had told him about a disease called heart block that happens when natural electrical impulses from the heart's upper chambers, or atria, fail to reach the lower

You're kidding: Frog stops people croaking

The pacemaker relies on a battery. The battery can thank a frog for its invention. In the 1780s, Luigi Galvani, an Italian anatomy professor, discovered that the body of a dead frog hanging on a copper hook in his laboratory twitched every time it touched an iron railing.

The twitch was caused by electricity flowing from the more reactive metal iron to the copper through the body fluids of the dead frog, making the muscles contract.

Shortly after Galvani's discovery, another Italian scientist, Alessandro Volta, built the world's first battery. It was a pile of alternatively stacked copper and zinc plates separated by paper moistened with salty water.

chambers, or ventricles. The disease can cause irregular heartbeats that can induce shortness of breath, and in bad cases loss of consciousness and even death. Wilson set out to solve the problem.

A recent photo of Wilson Greatbatch shows the inventor in a black tuxedo and crimson bowtie. He is clutching a massive circuit diagram of the implantable pacemaker. In his right hand are a couple of wires attached to red and black electrodes. The 80-year-old's hair is receding, his moustache grey, but there is nothing old about the shining eyes that beam at you through what could perhaps be the world's largest glasses. Wilson's strong, open, friendly face gives one a sense of the affa-

ble, persuasive nature that helped him establish and lead a highly successful research and development business.

A stroke of luck

The annals of discovery and invention are full of enthralling stories of accidents and mistakes that led to great breakthroughs. 'Dr Brainbox, was it years of painstaking research, trial and error, perseverance and sacrifice that resulted in your first producing WonderGel?'

'Um, well, one night I kind of, ah, dropped a beaker of monomaxovegemite into my beer. When I woke up next morning, there was WonderGel all over the laboratory floor.'

Working as an assistant professor of electrical engineering at the University of Buffalo, Wilson was using an oscillator to record heartbeats. A simple mistake showed him the way to make an implantable pacemaker.

Here's how it happened.

The oscillator needed a 10 000 ohm resistor. Wilson can't have been wearing his windscreensized spectacles because when he reached into his resistor box, he misread the colour coding and picked up a resistor a hundred times stronger. With the wrong resistor in place, the circuit generated very short, millisecond pulses followed by a one-second break during which the transistor was cut off. Staring at the circuit in wonder, Wilson realised immediately the value of his circuit – it would be perfect for driving a human heart.

On 7 May 1958, Wilson walked into the local Veterans Administration hospital with the world's first implantable cardiac pacemaker. He had constructed it using two Texas Instruments transistors.

In the hospital's animal lab, two surgeons exposed the heart of a dog and touched it with the two wires protruding from the pacemaker. The heart began to beat in time with the pacemaker's pulses.

Wilson recalled that the surgeons were delighted that his device worked. Seeing his little electronic pacemaker control a heart was one of Wilson's greatest ever thrills.

During the following two years, Wilson and his colleagues did more experiments with animals before they felt that the pacemaker was ready for a human patient. At the time of the first human implants, Wilson was working for a local electronic instrument company. The company supported his work but could not get the insurance necessary for implanting pacemakers into people. Wilson decided to start out on his own.

He established a licensing agreement with Medtronic, a medical electronics firm, in which they would make the pacemakers and Wilson's new company would manufacture the batteries.

Today, Wilson Greatbatch Ltd operates three highly successful large battery plants and research facilities in New York State.

With 600 000 of the devices implanted annually, about three million people worldwide owe their lives to pacemakers.

Wilson's philosophy: 'I think the world's concept of success and failure is sort of mixed up. I don't think the good Lord really cares whether you succeed or fail. He just wants you to try, and try hard.'

HOW IT WORKS:
The heart of the matter

Measuring about the size of your fist and weighing 250–300 grams, your heart pumps blood throughout your body's arteries, capillaries and veins. Right now, unless you're reading this on an exercise bike or, if you're adventurous, on a real bike, your heart is probably beating about 50 to 70 times a minute. If you're on that bike, your heart rate may be pumping at two or three times the resting rate.

An adult's heart pumps about 5 litres of blood each minute. That means a bucket of blood is pushed around your body every two minutes, and a beer-keg equivalent of the red stuff in ten minutes. That's a 4500 litre tank worth of blood in under a day. Hand us a calculator and dream up enough improbably large containers and we could keep going like this for paragraphs. Your heart pumps an Olympic-size swimming pool worth of plasma each year. So a swimming pool lifeguard's heart guards a swimming pool lifeguard's life by pumping a swimming pool volume of blood annually. Over a typical lifetime, your heart will pump about 200 000 million litres of blood.

Your heart comprises four pumps, each a chamber surrounded by muscles. These pumps must operate with perfect timing, sequentially filling and squirting, filling and squirting. The pattern is absolutely critical. Each muscle in this complex cardiac orchestra must play its part, contracting and expanding on cue in the correct progression. What ensures that never a beat is missed? Electricity. Yes, the cardiac orchestra relies on an electrical conductor.

The electrical signals that control heartbeat originate in a group of cells at the top of your heart called the sinoatrial node. The current travels rapidly down through your heart, triggering the chambers in turn. The cells of the sinoatrial node are called the pacemaker of the heart because the rate at which these cells transmit electricity dictates the rate at which the entire heart beats.

One of the ways that the sinoatrial node knows when to initiate a heartbeat is through signals from a sensor in the thick carotid artery in your neck. This artery expands as a pulse of blood passes through it, shrinking afterwards. Time for another pump, announces the sensor via electricity to the sinoatrial node.

When the body's natural heartbeat stimulator fails, an implanted pacemaker can generate the regular pulses to keep your heart ticking away. Even more cleverly, implanted pacemakers can sit back and do nothing except monitor the normal activity of the heart. If the natural prompt for a beat fails to occur, the pacemaker fires.

SUCCESS SCALE

INNOVATION

Wilson's pacemaker differed from existing ones in one key respect. It was smaller, which enabled it to be implanted in the human body. This was an incredible advantage, giving full mobility to people who need a pacemaker, and millions worldwide do.

MANAGEMENT

Good people and leadership skills helped Wilson succeed in business.

BUSINESS SKILLS

Others might have given up when they realised that it would not be practicable to manufacture and market their own invention. Wilson was astute enough to see opportunities in allied areas. After all, a pacemaker is nothing without a battery. Wilson's company created long-life batteries for pacemakers as well as pacemaker components.

References

Internet

<www.livingprimetime.com/AllCovers/dec1999/workdec1999/wilson_greatbatch_man_of_the_mil.htm>.

<www.circ.ahajournals.org/cgi/content/full/105/18/2136/FIG1>.

Great Moments in Science, 'Flatline & Defibrillator – Part 1, by Karl S. Kruszelnicki, at <www.abc.net.au/science/k2/moments/s1422463.htm>.

Harvard Medical School, Consumer Health Information – Pacemaker, at <www.intelihealth.com>.

John Hopps, who invented the pacemaker, at <www.ct.essortment.com/johnhoppswhoi_rlou.htm>.

Bubble, Toil and Trouble

How a small sphere enabled Guinness to tap into a global taste

Inventors will go a long way to protect their creation. Some use secrecy, others register a patent. Arthur Guinness chose a pick-axe.

Brewing was a risky business back in the 1700s. Arthur Guinness, born in 1725 in County Kildare, Ireland, was destined to brew the dark stout that today bears his name. Beer ran in Arthur's blood. His father, Richard, managed land for the Archbishop of Cashel and regularly brewed beer for the workers on the estate to keep them happy. At the age of 31 and with a £100 inheritance from the archbishop's will, Arthur followed in his father's footsteps and set up a small brewing company outside Dublin.

In 1759, Arthur rented a run-down brewery at St James's Gate, Dublin, which would one day become the world's largest brewery. With foresight and vision that would excite many an estate agent today, Arthur signed a 9000-year lease.

But in 1775, less than 0.2 per cent into the lease, the Dublin

Corporation demanded Arthur pay a levy for the clean spring water that flowed from the Wicklow Mountains south of Dublin. Arthur refused to pay, citing the terms of his £45 per year lease that included rights to draw water. The Dublin Corporation sent a group to St James's Gate to fill in the water-course. And that's where the pick-axe came in.

Arthur wielded the pick-axe and stood his ground, threatening the Dublin Corporation committee and attending sheriff with the foulest of language. Thankfully, nobody was hurt. The Dublin Corporation retreated and the dispute was finally settled in 1784 with the granting of water rights for another 8975 years.

The birth of black beer

It was this kind of maverick, no-nonsense approach that drove Arthur to take on the London breweries in producing porter, a dark beer that had been popular with Covent Garden porters since the 1730s. Arthur also demonstrated the single-minded dedication that we see in many of the entrepreneurs described in this book. In 1799, with Arthur's brewery having driven all English imports of the dark brew from the Irish market, the Guinness brewery ceased the production of ales to concentrate on what later became known as Irish or Guinness stout.

In 1886, producing 1.2 million barrels of stout a year, Guinness became the world's largest brewery. (You could probably check this fact in the *Guinness Book of World Records*, a publication the company launched in 1955 to settle trivia arguments in pubs and to promote the beer.) In 1906, 1 in 30 people in Dublin depended on the brewery for their livelihood.

While 10 per cent of Guinness sales were to overseas customers by 1870, it wasn't until 1963 that the first brewery opened outside Ireland and Britain – in Nigeria, which is the third largest and also the fastest growing Guinness market in the world.

True Guinness drinkers claimed that nothing tasted like the drink brewed in Dublin. So, rather than resting on the creation of an iconic drink in Dublin, Guinness set out, using science and innovation, to improve its taste overseas and to make a great product even greater.

Innovation in a can

Science had played a part in the Guinness brewery since the 1893 appointment of university science graduate T. B. Case. The first research lab was established in 1901, and an experimental brewhouse and maltings followed. But it was a technological achievement that appeared in a can of Guinness in 1988 that launched true-tasting Guinness around the beer-drinking world.

The first canned beer had been sold in 1935 in the USA by the Krueger Brewing Company of Virginia. But Guinness didn't pour well from cans due to the lack of high-pressure taps that were required to produce its distinctive, creamy head. Things changed in 1985, after 25 years of research costing around $12 million, when Guinness product developers Alan Forage and William Byrne filed a patent for the Guinness widget, which was introduced to the canned product.

In the past, the word 'widget' had been given to devices that had defied description. Luckily, similar terms, such as 'dinkus', 'doohickie', 'thingamajig' or 'gewgaw', were not given to the device inserted in our beer cans.

The word is probably derived from the word 'gadget', and was first used in 1924 in a play called *Beggar On Horseback*, the joke being in the fact that the audience never learns what a widget is, so our apologies for spoiling this hilarity if the play comes to a theatre near you.

The widget enabled people anywhere – at home, on a picnic, wherever – to enjoy the same thick, creamy head that traditionally could only be achieved outside Ireland in a pub, and usually an unsavoury Irish theme pub at that.

A toast to the future

So perfect Guinness could now be had in a pub or in a glass poured from a can. The next step was to produce Guinness that could be consumed straight from the bottle. The bottle widget apparently required several more million pounds of research and development and – we assume – tasting by Guinness's technical support and innovation team. The bottle widget, launched in 1999 and shaped like a rocket to ensure that it remains upright, works as the can widget does, enabling the beer to form a head when the bottle is opened. Then, with each swig of Guinness, the creamy head is refreshed to ensure a smooth consistency down to the last sip.

Bottled beer makes up 90 per cent of the American beer market, so the widget enables Guinness to compete in a market to which it previously had only about 10 per cent access. Prior to the invention of the widget, Guinness sold about 7 million pints a day around the world. Today, worldwide consumption of Guinness is more than 10 million pints each day.

In 2001, almost 2 billion pints of Guinness were sold around the world in more than 120 countries, and Guinness was brewed in 52 countries outside of Ireland. Back at St James's Gate in Dublin, more than 4 million pints of Guinness are still brewed each day. And thanks to Arthur's 9000-year lease, in the year 10759 Guinness may well still be brewed there on the original site.

Early days:
Keeping it in the family

Beer, which, after tea, is today the world's most popular drink, was brewed by the Babylonians more than 6000 years ago. The word 'beer' may originate from the Hebrew word for grain, the Saxon word for barley or the Latin word for drink. But the first known use of the word 'stout' (meaning full-bodied) was the Stout-Porter brewed by Guinness in 1820.

Throughout most of its history Guinness was truly a family business. Arthur died in 1803, leaving the company to his son Arthur Guinness II, who, on his death in 1850, handed the business to his son, who was to become Sir Benjamin, Lord Mayor of Dublin. On his death in 1868, Sir Benjamin handed the business on to his son, Arthur, who sold it to his brother Edward. Edward, his son, and then his grandson chaired the company into the 1980s, when the family presence on the company board declined. Although the family still has a large financial stake in the company, there has been no direct involvement in the management of Guinness or its holding company, Diageo PLC, since 1992.

You're kidding:
Guinness really is good for you

Comedian Henny Youngman once said, 'When I read about the evils of drinking, I gave up reading.'

The 'Guinness is good for you' advertising campaign was based on market research that showed people felt good after a pint of Guinness. The saying was first heard in 1928 and continued through the 1930s and 1940s.

John Gilroy, who developed the famous campaign at Benson's Advertising, may well have had medical results on which to base his campaign. In 1815, Guinness is said to have helped the recovery of a wounded cavalry officer at the Battle of Waterloo.

In the mid 1900s, postoperative patients and blood donors in Ireland and England were given Guinness due to its high iron content. Pregnant women and nursing mothers were also once advised to drink Guinness.

However, Guinness ceased the campaign decades ago and no longer makes health claims for the brew, to mothers or anyone else. Instead, they run advertisements that call for responsible drinking.

Nonetheless, results of testing in 2003 in the USA showed that the 'Guinness is good for you' campaign may have some truth in it still. Researchers at the University of Wisconsin found that dogs given a pint of stout with meals had fewer blood clots than dogs given lager. The researchers said that antioxidants in Guinness slowed the deposit of cholesterol on artery walls. The dogs, they noted, said woof with a slight slur.

Although non-Guinness drinkers may complain about it being more a meal than a drink, a pint of Guinness actually contains fewer calories than a similar-sized serving of milk or orange juice.

Guinness drinkers say that it is less a product than a state of mind, although the state of mind is probably just inebriation.

HOW IT WORKS:

Beer bubbles and widget workings

Most beers are fizzy because they are carbonated with carbon dioxide, which is dissolved in the beer under pressure. When the pressure drops, through opening a can or with removal of the beer from a keg, the carbon dioxide bubbles out of the beer, thereby forming the familiar head.

As well as the carbon dioxide, which is also dissolved in fizzy beers and canned drinks, Guinness – even in its pre-widget days – also contains nitrogen. With a mixture of carbon dioxide and nitrogen bubbles, it is less fizzy but its head lasts much longer. In pubs it is poured under high pressure through a tap with six fine holes, which creates a creamy head made up of tiny bubbles 0.1 millilitres across. Two-thirds of the glass is poured, left to settle and then topped up. Advertising campaigns claim it takes 119.5 seconds to pour the perfect Guinness pint.

Bubbles of nitrogen gas are smaller than carbon dioxide bubbles, and hence more stable and longer lasting. Once poured, the bubbles in a pint of Guinness appear to defy the laws of gravity by sinking, rather than rising as buoyant bubbles in a liquid would be expected to do.

Dr M. N. Hasan Khan at CSIRO Minerals modelled Guinness bubbles on a computer during his PhD on bubble behaviour at the University of New South Wales. After calculating the forces acting on bubbles, he found those near the sides of the glass would sink. He explained that bubbles rising in a central column in a pint of Guinness drag the liquid upwards. The bubbles then spread out at the top of the glass, forcing the liquid at the sides to descend. This drags small bubbles downwards, going with the flow but apparently against the laws of physics. Hasan says that the brewing and wine-making industries could use such computational modelling for better design of their processes and equipment.

Try watching bubbles defying the laws of physics by closely analysing a pint of Guinness yourself. But ensure that you make your observations with your first glass of Guinness, as subsequent observations may be unreliable.

Hasan published his results in 1999, but a sceptical scientific world remained unconvinced that bubbles could float downwards, believing it was simply an optical illusion. Then, in 2004, scientists at Edinburgh and Stanford Universities used high-speed video to record the movement and give definitive proof that bubbles in Guinness do indeed travel downwards due to drag at the side of the glass overpowering the tiny bubbles' buoyancy.

The researchers announced their results on St Patrick's Day.

As for the widget, it is a small, hollow plastic sphere that is inserted into the bottom of a can prior to being filled with the Guinness. The sphere has a 0.61 millilitre hole cut by laser – Alan and William experimented with different widths, but this seemed to work best.

The nitrogen in Guinness ensures smaller bubbles, but the presence of this gas means the Guinness contains less carbon dioxide, so if you poured a pre-widget Guinness straight from the can it would not froth as much. Hence the need for the smart sphere.

Here's how it works. A small amount of liquid nitrogen is added to the can of Guinness before it is sealed. The evaporation of the liquid nitrogen into a gas raises the internal pressure, forcing 10–15 millilitres of Guinness into the widget. When the can is opened it depressurises, so the high pressure in the sphere rapidly forces the small amount of Guinness out of the sphere's small opening, recreating the spray of beer from the tiny holes of a tap in a pub. The drop in pressure also releases the nitrogen that has been dissolved in the Guinness. The combined effect is the formation of millions of tiny bubbles. The beer must be poured immediately before the nitrogen escapes completely, creating a long-lasting, smooth head in which you can sign your name or draw a shamrock.

SUCCESS SCALE

MANAGEMENT

The Guinness company was managed by the Guinness family for generations.

MARKETING

From the entertaining cartoon posters claiming 'My Goodness, My Guinness' to the modern absurdist television campaigns depicting a fish riding a bicycle, Guinness advertisements have been famous around the world for decades.

INNOVATION

Scientific research and development have played a part in Guinness products since the 1800s. The widget made a good product better.

References

Internet

Guinness home page at <www.guinness.com>.
How Stuff Works at <www.home.howstuffworks.com>.
New Scientist online, at <www.newscientist.com>.
Answers.com <www.answers.com>
Encyclopaedia Britannica online, at <www.britannica.com>.

Raging Bull

An Asian syrup was just the tonic for an entrepreneur with energy

Styria is a province in southeastern Austria. It is a beautiful place, lying almost wholly within the Alps. Think magnificent peaks and valleys, dense pine forests, glaciers, rivers and lakes.

It was in 1944, towards the end of the Second World War, that Dietrich Mateschitz was born in the small Styrian village of St Marein im Mürztal. His parents were primary school teachers. Dietrich's father spent years in a prisoner-of-war camp after the end of the war, so Dietrich didn't meet him until he was 11.

Dietrich's academic record gave no hints to the astonishing success that lay ahead. After secondary school, he enrolled at the University for World Trade in Vienna.

Friends recall that Dietrich was a keen party-goer; it took him 10 years to graduate with a marketing degree.

'Life as a student is enjoyable,' he later reflected.

He was a charming man. Friends described him as funny, full of ambition and full of crazy ideas.

The 28 year old began his working life at Unilever, the international food, homecare and personal care manufacturer. Soon he was marketing dishwashing detergents and soap across Europe.

Before long, he became marketing director for Blendax, a company that produced toothpaste that claimed to protect gums, as well as skin creams and shampoo. The new position meant a lot of international travel.

Drink to success

A few long flights and time away from home brought Dietrich a vision of a bleak, boring life as a travelling salesman, spending night after night in bland hotels were he to continue with this way of life. He wanted more. As you would have gathered, the people featured in this book get more.

In Dietrich's case the 'more' came in the form of a premium-priced sparkling drink based on a popular Asian beverage.

Dietrich was confident that he could formulate the pick-me-up in a way that would appeal to Europeans. He asked an advertising agency friend to come up with a logo for his newly-named Red Bull and artwork for the can. Eventually, Dietrich had an image he was happy with: two identical muscular red bulls charging at each other, and a golden orb in the background. The slim cans would feature a blue and silver background, which contrasted strongly with the Red Bull red bulls. After numerous rejected slogans the German agency came up with, 'Red Bull – gives you wings'.

Licensing of Red Bull in Austria was problematic, since a number of its ingredients, such as taurine, had never before been used in European food products. It took three years and numerous scientific tests on the new brew before the Austrian government would permit the drink to be sold there.

Now that one hurdle had been overcome, another seemingly insurmountable one emerged. Consumers didn't like the product. 'People didn't believe the taste, the logo, the brand name,' recalls Dietrich. 'I'd never before experienced such a disaster.'

With the singlemindedness that marks successful entrepreneurs and madmen, Dietrich ignored the initial comments. He

Early days:
Drink ... Mateschitz

During regular visits to Bangkok, Thailand, as marketing director for Blendax, Dietrich took a liking to a popular syrupy tonic drink called Krating Daeng (Red Water Buffalo) sold in pharmacies as a revitalising agent. He said that it helped cure his jetlag. By chance, one of his Thai licensees, Chaleo Yoovidhya, had shares in the local company that made the drink.

Dietrich quickly realised that the little syrups, originally developed in Japan, did extremely well across Asia. Taisho Pharmaceuticals, the tonic drink developer, was one of Japan's most successful companies.

There was no mystery about the ingredients of Krating Daeng – they were clearly described on the can. Even better for Dietrich, there was no trademark or patent to protect the formula.

In 1984, the 40-year-old Dietrich Mateschitz resigned from Blendax to concentrate on establishing a company to introduce the high-energy drink to Europe. The new business had a capital of US$1 million, with Dietrich and Chaleo each investing $500 000. Dietrich and Chaleo each owned 49 per cent of the company, with the remaining 2 per cent owned by Chaleo's son Chalerm. By agreement, Dietrich would run the business.

Now it was time to modify the drink for Western palates. Light carbonation would help sell the product, Dietrich decided. Red Bull would appeal more to Westerners than Red Water Buffalo. And, while it would have certainly been a talking point among English-speaking consumers, Dietrich wisely decided against using his surname as the moniker for the new drink.

established an office in Fuschl, a tiny lakeside village just outside Salzburg. His faith in those little silver and blue cans never wavered. In time, helped by a truly innovative approach to publicity, the drink that gives you wings was flying off the shelves. In 1992, Dietrich introduced Red Bull to its first foreign market, Hungary. Britain and Germany soon followed.

Now, Red Bull is energising people in over 100 countries around the globe.

Early publicity in Britain included paying students to drive around in Minis and Volkswagen Beetles with huge Red Bull cans on top. This inexpensive buzz marketing was highly successful; by the year 2000 Red Bull's annual sales in Britain had rocketed to 200 million cans.

The sports connection

Today, promotion is primarily through the company's prominent association with sports such as surfing, kite sailing, parachuting, mountain biking and Formula 1 motor racing. 'We don't bring the product to the consumer, we bring consumers to the product,' says Dietrich.

Then there is the annual Flugtag, in which contestants build flying-machines and jump off a rampart into water. Remember the slogan, 'It gives you wings'?

Certainly, the sales appear to have wings. In its first year, 1987, Red Bull sales totalled $1 million. Two years later this figure had doubled. By 1995, sales were $88 million. This had increased an incredible tenfold by 2001.

During 2004, Red Bull sold almost two billion cans worldwide in 120 countries. Today,

> *You're kidding:*
> # No dicking around
>
> Just in case you were wondering, in its publicity material, Red Bull points out that the taurine in its drinks 'is a purely synthetic substance produced by pharmaceutical companies'. 'It does not', the company assures us, 'come from bulls' testicles or semen, as the myth would have it'.

Dietrich Mateschitz is worth well over $2 billion.

Tastes like ...

Pop open the distinctive blue and silver can. Pour the contents into a glass. Red Bull has a yellow appearance. You could call it the colour of apple juice. You could also call it the colour of urine. It just depends on your frame of mind. And Red Bull might be just the thing for your frame of mind – that is, if you believe everything that you read on a can:

Carbonated Taurine Drink. With Taurine. Vitalizes Body and Mind.

Red Bull® Energy Drink – especially developed for times of increased stress or strain:

Increases physical endurance

Improves and increases concentration and reaction speed

Improves vigilance Stimulates metabolism.

Vigilance. Good word, that. But an odd marketing proposition. Why would you pay well over twice the going rate for fizzy drinks in order to be more wary and watchful? But Red Bull doesn't want you to be too vigilant. There is a warning: *Usage – 2 cans max. daily.*

Red Bull is sweet, tangy and slightly bitter. Paul describes it as tasting a little like non-alcoholic apple cider. Others describe it as like a liquefied sherbet bomb or like fairy floss. A trawl through the internet finds Red Bull described as almost berry, anise, carbonated medicine and, particularly descriptively, like melted gummy bears mixed with cough syrup. There are also numerous less flattering descriptions.

The drink is certainly popular: 'It's a great drink. It's effective. Red Bull makes you feel energised and alert. It has a refreshing taste, a nice taste.' This is the opinion of Paul's sister-in-law Helen, who happened to be walking past while he was writing.

HOW IT WORKS:
The full Bull

Red Bull is very high in sugar, in levels almost identical to those found in Coca Cola. It contains B-complex vitamins, caffeine and taurine.

Taurine is an amino acid known to chemists and brainiacs as aminoethanesulfonic acid. It is produced naturally in humans and found in seafood and meat. The compound helps with fat absorption and has a detoxifying effect. Breast milk contains high concentrations of taurine, although Red Bull warns that its product is not recommended for children, or pregnant or lactating women.

The European Commission's Scientific Committee on Food has had a close look at Red Bull and other energy drinks. Here's what they said: 'For taurine and glucuronolactone, the Committee was unable to conclude that the safety-in-use of these constituents in the concentration ranges reported ... had been adequately established. Further studies would be required to establish upper safe levels for daily intake of taurine and glucuronolactone.' So the committee is saying that they don't know what the stuff does, but if drinking lots of it turns out to be harmful, don't blame them.

The committee did discover that rats fed large amounts of taurine reacted weirdly. They became anxious, irritable and sensitive to noise. Some even began munching their own paws. In fairness to Dietrich Mateschitz, the levels of taurine fed to the rats were the equivalent to you or us slurping up to 70 Red Bulls per day for 13 weeks. At that rate, you'd be anxious and irritable too, especially if you found yourself more than 10 metres away from a toilet.

Caffeine is a common ingredient of energy drinks. A 250 millilitre can of Red Bull contains 80 mg, the same amount of caffeine as there is in a strong cup of brewed coffee. Caffeine is a stimulant. Our friends on the European Commission's Scientific Committee recommend that we restrict regular daily caffeine intake to no more than 300 mg.

'High levels of caffeine can boost heart rate and blood pressure, causing palpitations,' says Dr Pat Kendall, a human nutrition specialist from Colorado State University. 'Mixing these drinks with alcohol further increases the risk of heart rhythm problems. If caffeine makes you jittery, the drinks may actually impair performance.'

Norway, Denmark and France have to date prohibited sales of Red Bull, and it has only just been approved for sale in Canada.

SUCCESS SCALE

INNOVATION
See marketing.

PRODUCT
See marketing.

MANAGEMENT
Outstanding marketer.

MARKETING
Yes.

THREATS
Better marketing.

References

Internet

Red Bull home page, at <www.redbull.com>.
'Opinion of the Scientific Committee on Food on Additional information on 'energy' drinks' 2003, at <www.europa.eu.int/comm/food/fs/sc/scf/out169_en.pdf>
Northern Illinois University Health News, at <www.star.niu.edu / health/q_and_a/articles/020703.php>

Text

Dolan, Kerry A. 2005, 'The Soda With Buzz', *Forbes Magazine*, 28 March.

Ripe for Success

Macpherson Robertson used 21st century marketing methods in the 1800s

A number 112 tram bears us from the city into nearby Fitzroy, where we travel along Brunswick Street, a busy shopping strip that is a mixture of commonplace and chic. There are old-fashioned barbers' shops next to trendy hairdressers' salons, laundrettes and Vietnamese supermarkets adjoin expensive restaurants. Here and there, especially if you look above the shopfronts, are vestiges of nineteenth century architecture.

It is nineteenth century architecture that we are pursuing today – or, more particularly, the surviving remnants of nineteenth century entrepreneurship.

We alight, negotiate the Brunswick Street traffic and walk east along narrow Argyle Street. The slender, tin-roofed, Victorian terrace houses, some featuring original ornate brick patterns, others painted cream, take us back in time. In the distance, a large, red-brick building is visible. That building would have dominated the lives of many of the former residents of these homes. Each morning, from Monday to Saturday, men and women would have stepped out of their front doors and set off for the factory. We retrace their steps.

The old building takes up the whole block at the end of Argyle Street. It is substantially built, two-storeys high, with ceiling-to-floor windows upstairs. The windows are a recent addition. Today the building serves as luxury apartments. But there are plenty of clues pointing to the true origins of this landmark. The first we spot is a sign saying MacRobertson Cl. A few metres away is MacRobertson La. Screwed high on a rendered wall is a small white plaque featuring the word 'MacRobertson's' rendered in grey in distinctive trademark swirl.

We walk down tiny MacRobertson Lane. Four metres above, steel girders link adjoining buildings. These girders would have borne much weight in their time – heavy equipment and massive amounts of sugar, milk, cocoa, butter and cream.

Around the corner is a far more imposing sign featuring the MacRobertson swirl. Of course, the ingredients listed above make chocolate. Macpherson Robertson was a man who built a fortune on chocolate.

Apprentice chocolatier

As a young man, Mac felt that he had the skills to do well in the confectionery business. At the time, Melbourne boasted six major manufacturers. *MacRobertson, The Chocolate King*, Jill Robertson's readable, well-researched biography, lists favourite products of the 1870s: there were conversation lollies, with words printed on them, peppermint lozenges, jubes, soft creamy fondants, toffees, boiled lollies and cream sweets.

Mac was offered an apprenticeship at which he earned 15 shillings a week. As he reflected years later in the *Australasian Confectioner and Soda Fountain Journal*, he felt he was 'ascending bit by bit the confectionary ladder'. A quick learner and a conscientious worker, within a few years a competing company appointed Mac as chief sugar boiler. What a shame that job titles now lack such endearing specificity. 'Senior technical officer' just doesn't have the same ring to it as 'chief sugar boiler'.

Despite the new title, Mac was keen to launch his own business. And so he did. He bought bags of sugar and boiled them up in

Early days:
From boiled lollies to chocolate

Macpherson Robertson was born in Ballarat, Victoria, in 1859 to an Irish mother, Margaret, and Scottish father, David. Mac was the first of seven children. These were goldrush days. David was convinced that he would find sufficient gold to support his family. Margaret, ever the realist, thought that the family's prospecting prospects were poor, so she encouraged her husband to work at carpentry and other jobs. The early years were difficult for Mac. Money was tight and the family frequently moved in search of a better life. When Mac was nine, Margaret took the family to Scotland so she could care for her ailing mother in-law. Meanwhile, Mac's father sought work in Fiji. Life had been grim for the family in Australia; in Scotland it was grimmer and freezing as well. This was real poverty; people were starving.

As the eldest son, Mac had to work to help support his family. He pursued various jobs, many of which were tedious and gruelling. Importantly for this story, he picked up a labouring position at a local confectionary factory. Rising at five each morning, Mac worked 13-hour days helping to make boiled lollies, barley sugars and jubes. The work was hard but Mac was fascinated by the magical processes that transformed raw ingredients into delectable, attractive sweets.

To outsiders, the processes may have well have seemed magical. Confectioners guarded their secrets. There was no book that would tell you the correct temperature at which molten glucose should be poured and mixed with other ingredients to make the perfect lolly. You learnt this and a hundred other facts on the job, if you were lucky, and if you had the interest and intelligence to remember it all.

Following the death of Mac's grandmother, Margaret brought the family back to Melbourne. The Robertson family never did profit from the goldrush, but Victoria certainly did. These were heady times, in which the wealthy colony was rapidly growing and expanding. Perfect time for young Mac to arrive on the scene.

the bathroom of the family's Fitzroy home.

Going out on his own also meant going out on his own, literally. Mac had to take his products to the people. Every day he pounded the pavements of Melbourne carrying his wares. Hard work, Mac's captivating manner and the quality of the home-produced products proved a successful combination. Mac soon needed to expand his 'factory', which he did by buying two adjoining properties.

Macpherson Robertson needed a catchy name for his business. It didn't take him long to link his abbreviated given name with his family name to come up with MacRobertson. He wrote his name with a flourish in a form still distinctive today.

Mac makes his mark

By 1886, the 27-year-old Mac was employing thirty people. Needing more space, he bought a building at the end of the street, demolished it and constructed the two-storey factory that we so admired on our walk.

Mac was a canny businessman. As *MacRobertson, The Chocolate King* points out, Mac saw his future depending on two things: using the most modern manufacturing techniques available and capturing the public imagination with innovative, high-quality goods that sold at a reasonable price. As many have done since, Mac decided to see what he could learn for his business from techniques and trends overseas.

Around the world he went, returning five months later full of ideas for new products to tempt Australia's tastebuds. One of the products was chewing gum, which was very popular in America. Mac made it in flavours that included peppermint, liquorice and aniseed and clove. He also introduced fairy floss to Australia.

In the 1890s, as well as the more traditional sweets and the gum, MacRobertson's produced chocolates, cocoa, jam, jellies and tomato sauce. You could also buy treacle balls, silver sticks, honey balls, heliotrope lozenges, fig toffee and moonshine biffs, squares of marshmallow rolled in toasted shredded coconut, what Australians know as toasted marshmallows.

Mac now employed many staff. In today's terminology, Mac was an excellent people person. He was a fine leader, knew all his staff by name and regularly walked around the factory floor. MacRobertson's had a reputation as a friendly place in which to work. Mac paid as much attention to overseeing the quality of the products as he did to initiating a wide range of advertising and marketing initiatives. There were posters, postcards, booklets and competitions. The trademark delivery wagons, drawn by handsome thoroughbreds, were a common sight around Melbourne.

A splendid publicity photograph of the 39-year-old Mac shows him in profile, a look that he hoped would help charm the voters in his bid for local government. He is a handsome, strong-looking man, with dark, wavy hair, short at the back and sides, with immaculately greased curls crowning his forehead. A luxuriant moustache caps his mouth; his strong chin is held high. Mac also wears an expensive-looking jacket with high lapels. A wide, plain tie, with an ostentatious pin through the knot, holds together a stiff white collar. Intently, his eyes stare ahead. Here is a man who knows where he's going. In this instance, though, the local voters decided that he could go elsewhere; he lost the election.

Growing vertically

With numerous factories throughout Fitzroy, Mac Robertson decided that he needed a guaranteed supply of ingredients, especially glucose, which he was importing. So he established a new company in Melbourne to manufacture the glucose from maize supplied by another Mac enterprise, this one in Queensland. The glucose was stored in wooden casks. A cask shortage prompted Mac to move into the timber business. He presented a perfect example of what marketing gurus would come to refer to as 'vertical integration', expanding into different points along the production path.

Mac's visionary attitude and lifelong dedication to business helped to make Macpherson Robertson one of the country's wealthiest people. Australians' sweet teeth and the nation's growing population would have helped, but Mac succeeded in an industry where many failed. By the late 1920s, he employed 2500 staff and owned nineteen buildings. A decade later his staff numbered 5000.

In 1924 MacRobertson's launched a new chocolate bar that is to this day one of Australia's most popular. A combination of cherries and coconut wrapped in dark chocolate, Cherry Ripe came in distinctive red wrappers, emblazoned with the phrase, 'the Big Cherry Taste'.

You're kidding: Big Mac

Mac was a great philanthropist whose name is still prominent throughout Melbourne. The projects he funded all bear the MacRobertson name; if they led to extra product sales, so much the better. MacRobertson's Girls' High School has educated thousands of Melbourne girls. There is the MacRobertson Bridge over the Yarra River, the MacRobertson Fountain behind the Shrine of Remembrance and the MacRobertson Herbarium at the Botanic Gardens.

There is even MacRobertson Land in Antarctica, the name bestowed by Douglas Mawson, grateful for the financial support that Robertson gave the 1929–30 British, Australian and New Zealand Antarctic research expedition.

Six years later, MacRobertson's created another famous product. It was a small chocolate frog that sold for a penny. Twopence would buy you a peppermint, malted milk or honey-flavoured, cream-filled version. Packing department foreman Fred McLean had his name appropriated for the smiling, large-eyed Freddo frogs.

Macpherson Robertson died in 1945. MacRobertson's continued until 1967, when the multinational Cadbury took over the company that Mac had founded 87 years earlier.

The last word goes to Leslie Cranbourne, former MacRobertson's production manager. Quoted in *MacRobertson, The Chocolate King*, he fondly recalls his former boss: 'He was a very fine man. He never got angry with anybody. In fact, he never

did a lot of talking. I think the main reason he was such a success was that everyone respected him – he knew the business, he knew everyone's name, he was fair and he had this mixture of power and humility.'

Marketing Mac

How does MacRobertson's approach to business look to a twenty-first century marketing expert? Over a coffee, Paul spoke to Michael Dowling, strategic account manager with Achieve Global, a company that trains business people. Michael has years of marketing experience.

'Mac is employing tactics that we'd use today,' explained Michael. 'Mac decided he wanted to create a brand with a whole lot of values associated with it. The brand has to give customers a satisfying experience.'

Today, the conventional approach is that you don't even talk to people about your commercial idea until you have prepared a business plan. Where was Mac's plan? 'The business plan he had was probably in his mind, although it may have been written down. He certainly had a clear vision of where he wanted to go,' said Michael.

HOW IT WORKS:
The food of the gods

Chocolate comes from a tropical plant that the Central American Mayan Indians named as *Theobromo cacao*, the food of the gods. Far more mundanely, we just call it the cocoa tree.

The cocoa tree produces purple or creamy-white beans in leathery pods. Throw those bitter beans into a bin. That's the first stage in chocolate making. Easy, eh? Over a week or so, bacteria transform the beans into a rich brown colour. They may even have a faint chocolate aroma.

The next step is to roast the beans and discard the shells. The kernels are crushed to make chocolate liquor paste, which solidifies to form bitter baking chocolate. Alternatively, the chocolate liquor is pressed to reduce the fatty yellow cocoa butter content and transformed into cocoa powder. A final option, which wins our vote, is to combine the liquor with sugar and additional cocoa butter. The result: sweet, eating chocolate.

To make milk chocolate, you add milk. You're probably getting the hang of this now.

Kneading chocolate for hours or days makes it smooth. Varying the temperature imparts particular flavours. Finally, the chocolate is strongly heated, which reduces the size of the fat crystals and produces that characteristic velvet quality.

Despite popular belief, there is little naturally occurring caffeine in chocolate. The myth might have come from the fact that chocolate contains a little theobromine. This chemical is an alkaloid, a notorious family that includes morphine, strychnine, ephedrine and nicotine. For some people, theobromine has similar effects to caffeine – alertness, elevated mood, reduced appetite, and increased mental and physical energy. Eat too much chocolate and you may experience insomnia and tremors, as well as a growing familiarity with dentists and a loathing for bathroom scales.

'He understood the importance of diversification, travelling the world looking for ways in which to be innovative.'

'A business plan is a communication tool. It sets out the vision, how to get there, the short- and longer-term goals. It forms the basis for the marketing elements.' The plan, according to Michael, should be a living document. 'You should be able to write your business plan, at top level, over two to three pages, no more. Otherwise, it just ends up being a doorstop.'

Was Mac plain lucky?

One question that we keep asking ourselves is, How much does luck play a part in helping people be successful? Would MacRobertson have been successful in other pursuits or did he simply stumble on a great opportunity? How hard could it be to sell people chocolates and lollies?

Michael leant back in his chair and looked thoughtful. 'Mac had the drive, the vision, marketing and people skills. He knew what he wanted to achieve. He had the managerial and leadership capability to make it happen. He would always have been successful in pretty much any industry.'

MacRobertson engaged well with his staff – he was famous for it. Couldn't he have done just as well without this? 'Probably not,' answered Michael. 'He got the most out of staff. There was a culture that was respectful with minimal work conflicts – we're all in it together.'

SUCCESS SCALE

INNOVATION
Mac took control of many stages of the production process, from sugar growing to making casks. There was profit to be made at each step.

MANAGEMENT
Energetic and imaginative. In an era before managers who know only how to manage, Mac had an intimate knowledge of his business.

MARKETING
Well ahead of his time. Mac knew the power of branding and the value of building a powerful profile.

References

Internet

Cadbury home page, at <www.cadbury.com.au/history/>.
Macpherson Robertson, at
<www.whitehat.com.au/Melbourne/People/MacRobertson.asp>.

Texts

Robertson, J., 2004, *MacRobertson: The Chocolate King*,
Lothian Books, Melbourne.
Encyclopaedia Britannica online, at <www.britannica.com>.

Give Credit to a Great Invention

Frank McNamara earned a lot of credit when he joined the millionaires club

The year is 1949. The location is midtown Manhattan: number 35 West 33rd Street, to be precise. It is early evening, and the shadows are lengthening. No shadow is longer than that of the Empire State building, then the world's tallest building, just metres away. The 102-storey monolith points heavenwards, its striking vertical lines capped by a dark, finely-pointed spire. This is the world's tallest building.

It is fitting that number 35 West 33rd Street is so close to the icon of mid 1900s free enterprise, for this is to be the location of one of the great developments of capitalism.

Businessmen in long coats and full-cut trousers stride along the pavement, jostling past women wearing short jackets or cardigans and knee-length skirts. There is a man wearing a maroon jacket and a garish tie featuring rodeo riders and bucking broncos. We will let him pass, for alighting from a slightly battered New York cab is the star of this story – Frank X. McNamara. Frank is a genial-looking man. His hair is black and carefully slicked back; it has a parting as straight as the spire up the road. A small, finely-trimmed moustache completes the picture. He wears a conservative dark suit and a white shirt and multihued tie.

Clothes are key to this tale, as you – and Frank – will shortly discover.

Frank McNamara is head of the Hamilton Credit Corporation, and gnawing at the head's head is a problem. One of Frank's customers has borrowed money to cover debts. These debts arose when the man lent neighbours his department store and petrol station charge cards to help them during an emergency. Now, says the man, the neighbours will be unable to pay him back for quite some time.

A problem shared is a problem solved, and this evening Frank has two people with him. Department store owner Alfred Bloomingdale is a long-time friend. Ralph Schneider is Frank's lawyer. The three have gathered at number 35 West 33rd Street because this is the location of one of New York's best known restaurants, Major's Cabin Grill.

Over steaks and martinis they discuss many things. One is, of course, the wayward Credit Corporation customer.

Invention on the cards

Now, the meal is complete. Frank reaches into his pocket for his wallet. It is not there! He fumbles through all his pockets before recalling that he changed suits before setting out for dinner. His wallet is in his other suit. Frank has no cash with him. Highly embarrassed, he asks the maitre d' if he can use the restaurant's phone. He rings his wife, explaining the predicament. She hunts out his wallet and brings it to West 33rd Street so that her husband can pay the bill. Frank declares that he will never again be caught short at a restaurant.

That evening he begins to think about what happened. Why should people be limited to spending what they are carrying in cash, instead of being able to spend what they can afford? he wonders.

He has an idea, which he explains to Ralph Schneider. Frank's idea was the birth of what we today know as credit cards.

Buy today, pay tomorrow

Soon after that fateful dinner, in February 1950, Frank and Ralph return to Major's Cabin Grill and enjoy a sumptuous meal. When the bill comes, Frank declines to pay in cash. Instead, he presents a small, cardboard card and signs for the meal. The restaurant cheerfully accepts Frank's request for credit. Keeping the restaurant cheerful was the fact that Frank was an affluent, well-known diner.

There was nothing new about the idea of buying today and paying tomorrow. For hundreds of years, clothing merchants had given customers garments in return for small weekly payments. Some merchants used a wooden stick to keep records of people's purchases. Notches on one side of the stick showed the amount owed to the merchant. The other side of the stick had a notch representing each payment.

In the USA in the 1920s, large stores, hotel chains and oil companies introduced a system of credit. They presented trusted customers with a small card. Instead of paying cash, the customer would show their card after a purchase. Every month or so, the customer would receive a bill for their purchases.

There is no point in inventing something that has been around for years. Frank McNamara set his mind to coming up with a card that allowed credit at a number of restaurants. Major's Cabin Grill was good, but it wasn't that good.

Dining out

Frank and Ralph managed to sweet talk the proprietor of Major's Cabin Grill and the owners of thirteen other New York restaurants into accepting their new card. The name of the card? 'Diners', of course. The two Diners Club founders offered the card to 200 people, most of whom were friends or acquaintances. The first cards featured a simple design, with the names of accepting restaurants printed on the back.

The two men rented space on the twenty-fourth floor of the Empire State Building. From there, they could look down on the restaurant in which the busi-

ness idea had been born. Diners Club rapidly expanded. They ascended to the thirty-second floor, and then the seventy-seventh.

The second edition of the *Diners Club News* ('Published every month in the interests of the now over 22 000 members of the Diners Club'), published in November 1950, proudly boasted that:

> You can now say 'Charge it' in New York, Chicago, Boston, Philadelphia, Los Angeles and other cities from coast to coast. The Diners Club, now eight months old, has grown from strictly a local credit system for clubs and restaurants in the New York area to one of America's largest overall charge account setups.

Diners Club was the first universal card that could be used at a variety of establishments. Soon, more than 330 American businesses were accepting it. Joining 'the Club' cost you $3 per year and you had to pay back your full debt each month. Diners Club charged the businesses that accepted their cards 7 per cent of each transaction. The businesses didn't like losing 7 per cent of their income, but they liked even less the idea that Diners Club members would go elsewhere if they couldn't use their cards.

By the end of 1951, Diners Club had cycled more than $1 million through its cards. The company was now highly profitable.

But Frank McNamara doubted that Diners Club had a long-term future. Nobody, not even Frank, could foresee the multi-billion dollar credit card industry that we have today. So, in 1952, he sold his share of the company to his two partners for about $200 000.

It was in 1990, *Life* magazine declared Frank McNamara one of the one hundred most influential Americans of the twentieth century. His was perhaps the ultimate of all inventions – a little

You're kidding:
Right royal dosh

Today, there is an almost endless range of credit cards to choose from. If you elect instead to pay cash, there's an almost endless range of slang to choose from to describe the dough, dosh, moolah or bucks that you are handing over.

There is rhino, readies, spondulicks, quid, bacon, bread, cabbage, lettuce, mazuma and yellowback. Each nationality has its special terms for currency or specific denominations. In pre-1960s Australia, a bob was slang for a shilling, equivalent today to ten cents.

Originating in the USA, a grand is a well-known term for $1000. Pony is an old British word denoting £25. In rhyming slang, macaroni also means £25. A more convoluted piece of rhyming slang in 1870s Britain gave rise to the name 'Oxford' for a five shilling coin. The coin was known as a dollar, for which the rhyming slang was Oxford scholar. A monkey means £500.

When Australia decided to introduce decimal currency, there was great debate over what name should be given to the new currency unit. The prime minister of the day, Sir Robert Menzies, a committed Anglophile, proposed the 'royal'. Over 1000 submissions for a currency name were received, with the government eventually deciding, of course, on the dollar.

HOW IT WORKS:
The attraction of a magnetic strip

A credit card's black, magnetic strip works in much the same way as a length of magnetic recording tape on an audio or video cassette. The data on the strip are written by moving it past a small electromagnet. Pulses of electricity flow through the electromagnet into the card strip. This magnetises sections of the strip in a particular pattern. Card readers, such as those at supermarket checkouts and in bank ATMs, work in the reverse way. Dragging the card across the sensor in the reader sets up small electrical currents in a pattern that corresponds to the magnetic areas on the strip. These patterns form numbers and letters.

Magnetised into your card's strip are your account number and your name. There may also be your secret four-digit PIN in coded form, a code representing the country that issued the card, currency units and other information about your account.

Keep your credit card away from magnets – they can easily erase data on the strip.

card that can get you pretty much whatever you want. Payment? Worry about that next month.

Today, 30 million businesses worldwide accept Diners Club cards. In the USA alone, around 80 per cent of families own at least one credit card, with the average card-carrying household having six bank credit cards. Total credit card debt in the USA is over US$600 billion.

Around 69 per cent of Australians have a credit or charge account. The country's credit and charge card debt is currently more than A$30 billion.

As Charles Farrar Browne, a nineteenth century American humorist commented: 'Let us all be happy, and live within our means, even if we have to borrow the money to do it with.'

SUCCESS SCALE

INNOVATION

An original application of an old principle, buying on credit. Development of computers allowed efficient tracking of people's purchases.

MANAGEMENT

Frank McNamara's professional background was perfect for the new enterprise. It was a shame for him that he didn't predict the future growth of credit cards.

MARKETING

A versatile system that allows you to buy today and pay tomorrow was always going to be a winner. When people saw the convenience, they just had to have a credit card.

References

Internet

Frank X. MacNamara, at <www.dinersme.com/dc_fifty_content.htm>.

About.com, at <www.history1900s.about.com/od/1950s/a/firstcredit-card_2.htm>.

Diners Club International home page, at <www.dinersclub.com.au>.

Encyclopaedia Britannica online, 2001, at <www.britannica.com>.

Squares Learn How to Do the Twist

Mathematician + cubic puzzle = craze x millions

A man in his late twenties stares at the Danube river that flows beside him. It's not the blue Danube, because its meanderings though polluted Budapest in mid 1970s Hungary gives it more of a dark brown hue. He looks lonely as he stands there, smoking cheap Hungarian cigarettes and dressed in what could be mistaken for beggar's clothing. But this introvert named Ernö is a genius, and within five years the lonely, badly-dressed man will become the most famous person in Hungary, the country's first millionaire, and will initiate a craze that will touch one-eighth of the world's population.

Ernö likes puzzles and, on this peaceful summer's day in 1974, he is puzzling over an invention of his that keeps breaking. As he looks across the river on which he regularly kayaks alone, he notices the round pebbles on the river bank. Ernö observes that their sharp edges have been rubbed smooth by the combined efforts of time and the Danube's waters. The result is a collection of beautiful objects that move effortlessly against each other. In a moment of inspiration, Ernö Rubik realises this simple beauty could be the basis for the internal mechanism of his invention, a collection of coloured squares that, in 1980, will become a household phrase – and be described in the *Oxford English Dictionary* as Rubik's Cube.

What a square

When he's not wandering around alone on the banks of the Danube, Ernö Rubik gives lectures in interior design at the Academy of Applied Arts and Design in Budapest. As well as being an architectural engineer, he is a designer and sculptor. He teaches his students about geometry and three-dimensional forms using models he makes from paper, cardboard, wood or plastic.

Ernö was born in 1941 in the air raid shelter of a Budapest hospital during the Second World War. His mother was a poet and his father an aircraft engineer who built gliders; they divorced while he was studying sculpture at the Budapest Technical University. Ernö went on to study architecture at the Academy of Applied Arts and Design, where he remained as a lecturer.

It was in his job as professor, which paid a couple of hundred dollars a month, that he began tinkering with the geometric models that would make him probably the richest man in Hungary.

In his room at his mother's apartment, he stacked together twenty-seven small blocks of wood to form a cube, threading rubber bands through each of them to hold his invention together. Although he was particularly interested in the small blocks' ability to move around without falling apart, he noticed the way the cube would produce different patterns. But the rubber band snapped, and neither bands, nor magnets nor teeth and grooves could cope with the intensive manoeuvring of the individual blocks.

Colourful cube

Some time after noticing the way those pebbles on the banks of the Danube glided past each other, Ernö managed to perfect a cylindrical mechanism to hold his cube together. With just the right looseness to allow movement

Early days:
A puzzled pre-history

Rubik's Cube appears unique and original, but its roots lie in century-old traditions and timeless games. Amazingly, two almost identical cubes were patented independently in the USA and Japan at about the same time that Rubik was patenting his cube in Hungary.

A cube was considered by the ancient Greeks to be one of the five Platonic solids, special shapes with identical sides. In ancient China there existed a puzzle now called the Tangram. It is made of five triangles, a square and a parallelogram, but the pieces can be put together and transformed into a variety of different shapes.

Sam Lloyd's puzzle is well known by sight, but not by name – the 15 Puzzle. Invented in the late 1870s in the USA, it is a sort of two-dimensional Rubik's Cube, made up of fifteen interlocking tiles. The flat tiles slide past each other but cannot be removed from their flat, square frame. They display a picture or consecutive numbers, and the aim of the game is to slide the tiles from a state of chaos back to their original picture or numerical order.

The 15 Puzzle may have inspired Ernö, as he had played with one as a boy. But in a strange coincidence, while he was applying for a patent for his invention in Hungary, an ironworks owner in Japan named Terutoshi Ishigi came up with a similar cube to Rubik's, obtaining a Japanese patent in 1976. It was even held together with a similar internal mechanism.

Meanwhile, Larry Nichols in the USA had actually patented a cube before Ernö. However, his cube was held together by magnets and was rejected by every toy company he approached, including the Ideal Toy Company that accepted Rubik's product. A USA federal judge ruled, once the cube craze had died down in 1984, that the distribution of Rubik's Cube in the USA infringed Larry's patent.

and the right tightness to keep it all together, the cube kept its shape without any rubber bands, magnets or other lacing. Ernö experimented with different patterns on his cube's faces, including numbered squares and various symbols. However, nothing worked more majestically than the simplicity of six bright colours on the fifty-four small faces of the cube.

While the construction of the colourful cube was amazing enough, it was its deconstruction to a state of coloured chaos that was the real magic. Late in 1974, Ernö became obsessed with trying to return the cube to its original pattern, not knowing whether this indeed would be possible without retracing each of his previous moves. After a month of intensive effort, he solved the puzzle and realised he had an interesting game on his hands.

Ernö was amazed that a puzzle like this didn't already exist. He applied for a Hungarian patent for his cube in early 1975, which was approved two years later. Called the Magic Cube (in Hungarian, Bűvös Kocka), it appeared in Budapest toy shops in the final months of 1977.

Powerful puzzle

During 1978, the Magic Cube's popularity grew throughout Hungary. But it would receive only academic interest outside Hungary during the next year, despite being available by mail order in the United Kingdom from the Pentangle company. The cube's real success would only come about thanks to some clever marketing and dogged determination from two men who arrived on the scene in 1978.

Hungarian-born Dr Tibor Laczi was returning from Vienna where he had been on a sales trip for his Austrian computer company when he noticed a waiter playing with the cube-shaped toy in a small-town cafe. The waiter couldn't use the cube so he sold it to Tibor, a keen mathematician, for a dollar. Although interest in

You're kidding:
Maths, cubes and group theory

Rubik's Cube has an incredible 43 quintillion possible combinations – 43 252 003 274 489 856 000 to be exact. Only one of these configurations is the correct solution.

It would take 1400 billion years to rotate the cube through each of its configurations (the universe itself is only 14 billion years old). Put another way, if each person on the planet randomly rotated a Rubik's Cube each second, one would appear in its solved state only once every 300 years.

The shortest way to move between a state of complete chaos and the solved cube has been calculated by computer. It is known as God's Algorithm, because it describes the minimum number of moves it would take an all-knowing being to unscramble the cube. The fewest number of moves was originally calculated to be about fifty, but could be as low as twenty-two.

Such calculations are based on group theory, a branch of abstract mathematics developed in the 1800s. Group theory describes the behaviour of collections of related elements, such as the cube's 43 quintillion arrangements. Rubik's Cube helped visualise such abstract concepts, as well as other difficult-to-imagine scientific challenges, such as the behaviour of quarks and other subatomic particles.

The puzzle can now be solved in less than seventeen seconds by people who don't have much else to do with their time.

the cube had started to wane in Hungary, Tibor obtained Ernö's permission to take it to Germany's Nuremberg Toy Fair in February 1979 to attempt to popularise it in other countries. While there he met Hungarian-born Tom Kremer from Britain, who was similarly captivated by the cube. They decided to take the cube to the Western world.

Initially, Tom experienced the same challenge that many new ideas face in a global market: no large players were willing to invest. The cube was too difficult and expensive to make, it would be difficult to demonstrate its addictive nature in a television advertisement and it was seen as too geeky and academic. By not talking or crying, shooting or starring in a movie, it broke all the rules of the toy industry and it didn't even require batteries.

Success at last

Eventually, their persistence paid off, and in September 1979 Tibor and Tom managed to convince the American Ideal Toy Company that the cube was one of the wonders of the world. The company purchased one million cubes from the Hungarian manufacturer.

The cube's international success was assisted by Ernö's demonstrations – he could by now solve the puzzle of the cube in two or three minutes, which he did at toy fairs around the world in January and February 1980. By May, the first Magic Cubes were being exported from Hungary. Ideal Toy wisely avoided the plan to rename them the Gordian Knot, and came up with the name Rubik's Cube for the mass market.

Twisting around the world

The Cube craze spread around the Western world in the 1980s as fast as a celebrity rumour would today. In its first two years, more than 100 million official Rubik's Cubes were sold; another 50 million imitations were also sold. Even with Ernö reportedly receiving only about 5 per cent of the cube's proceeds, it is easy to see how, in his early thirties, he became the first legal millionaire in the communist bloc. Before its popularity died out in 1983, 250 million cubes had been sold and one in every three homes in the Western world had one. It became the fastest-selling toy on Earth and probably the most popular puzzle ever. In the twenty-first century sales have begun to increase again, edging towards a million sales in the USA in 2004. Suddenly it's cool to cube again, with the product achieving iconic status among a generation that wasn't even born when it was invented.

But it hasn't been all good news. Believe it or not, new medical conditions appeared, such as Cubist's Thumb and Rubik's Wrist. The cube even caused the breakup of relationships – as one married woman in Germany complained, her husband was so obsessed with his cube that he no longer spoke to her and in bed was too exhausted from playing with his cube to even give her a cuddle.

Good things come in threes

Ernö philosophically explains the obsession with the three layers of three blocks – he says the number three seems to have important meaning in nature, reflected in other relationships such as mother, father and child; heaven, hell and earth; or birth, life and death.

Ernö also explains the popularity of the cube as being due to

it being an imitation of the problem-solving required in everyday life. In fact, he even says that the cube is an improvement on life – it has a clear solution that can be found independently of anyone else. We think these explanations go a fair way to explaining why Ernö was a lone kayaker.

A smashing solution

For cube dropouts who were frustrated by their inability to solve the puzzle independently, assistance came in the form of several dozen books on the subject. For the really desperate, help came from companies capitalising on the cube's success by producing toys such as The Cube Smasher. Others, including Paul's then 5-year-old daughter Kate, promoted the technique of peeling off the small coloured stickers to leave a completely black cube, thereby achieving the solution of returning the cube to its 'original' state, having a single colour on each face.

Simon favoured the dismantling method, in which a twist in the correct direction would pop the cube so that its twenty-odd component pieces would scatter onto the floor. Not only would this allow him to put the cube back together in its correct order, but it also provided an insight into the beautiful mechanism that Ernö envisaged as he stood by the Danube in the 1970s.

HOW IT WORKS:
Cube and cubelets

For those who weren't around in the early 1980s, or who happened to be in a deep coma, a Rubik's Cube appears to be a simple cube made up of twenty-seven smaller cubelets. The internal workings are actually made up of six central faces joined to spring-loaded spindles, eight 3-sided corner pieces with curved internal flanges and twelve 2-sided edge pieces with rounded internal cutouts. (For those who can add up, that's only 26 cubelets – there was no twenty-seventh cubelet at the centre of the cube itself.)

SUCCESS SCALE

MARKETING

With the cube's addictive nature not able to be portrayed in an advertisement, demonstrating it at fairs was a winning idea. Having someone such as Ernö who could actually solve the cube in a few minutes must have helped.

MANAGEMENT

In the early days, the cube was a victim of its own success, with problems caused by Hungarian production being unable to keep up with international demand.

INNOVATION

What seemed like an innovative idea was, coincidentally, imagined by two other inventors at around the same time. But Ernö's unique innovation was in the internal mechanism that enabled the cube to be twisted by anyone.

References

Internet

Ernö Rubik, at <www.inventors.about.com>.
Rubik's Cube, at <www.rubiks.com>.
Encyclopaedia Britannica online, at <www.britannica.com>.

Texts

Discover, March 1986.

Bored? Board game?

How the pursuit of knowledge generated a new category of cerebral pastime

Name the recreational activity pursued by Scott Abbott and Chris Haney?

On a cold wintry Saturday, 15 December 1979, two friends, Scott Abbott and Chris Haney, were comparing their skills at various board games. Both men worked for Canadian newspapers. Scott was sports editor for the *Canadian Press*, Chris, photo editor for the *Montreal Gazette*.

They soon decided that it might be fun – and profitable – to come up with an idea for a new game. By nightfall, they had a concept. However, turning that concept into reality took a harrowing two years. It was well worthwhile though. Their game, Trivial Pursuit, has now sold more than 70 million units. Dubbed 'the biggest phenomenon in game history' by *Time* magazine, Trivial Pursuit is a highly social pastime, in which contestants compete to discover who has the greatest store of often-useless information. You might routinely forget where you have left your mobile phone, but your brain can redeem itself by naming the second person to set foot on the Moon.

In late 1979, the big question for Scott and Chris was this: How are we going to get enough money to develop our idea and make prototypes? They recruited Scott's brother John and a friend of his, Ed Werner, a lawyer. The four set up a company, tirelessly developing their game for more than two years. When their money ran out, they convinced friends and relatives to buy shares to raise much needed additional capital.

They offered Michael Wurstlin, an unemployed 18-year-old artist, five shares in exchange for developing design and artwork. By the time the first 1100 sets were ready, expenses had risen to $80 000. This meant that each set had cost $75 to make. No one would pay that much for a board game, certainly not one that was completely unknown.

In the end, the first games wholesaled for $15 so that Canadian retailers could charge $29.95. This was still very expensive for a board game in 1981.

Now it was time to showcase Trivial Pursuit at the Montreal and New York toy fairs. To the creators' horror, they sold fewer than 400 sets. Two major game companies turned them down.

Chris Haney recalls the grim period. 'At that point we could have been had for a song.'

What was the theme of the second edition of Trivial Pursuit?

Luckily, the creators' compatriots finally came to the rescue. Canadian stores sold out of the first edition and requested more supplies. In response, the team came up with a new edition, Silver Screen, which probed movie knowledge. This second edition cost $75 000 to prepare, but the 20 000 games all sold.

Local interest in Trivial Pursuit prompted Chieftain Products, a Canadian company that distributed games from Selchow and Righter in the USA, to send copies to three senior executives of the American company. They played a few rounds and loved the game. 'Let's try to knock off video games such as Pac Man and Donkey Kong', they decided. This was a brave decision. They were pitching a static, old-fashioned board game against the latest craze – guiding colourful little electronic critters around an obstacle course, with a music track and beeps for extra fun. Could having to name the colour of lobster blood, for example, really compete?

A public relations consultant mailed off a promotion to 1800 buyers attending the 1983 New York Toy Fair. For good measure, the consultant also sent the promotion to a bevy of Hollywood stars. Sales boomed.

By the end of 1983, American customers had bought 1 million games. Canadians supported their home-grown product even more strongly – with more than 2 million purchases. Selchow and Righter struggled to keep up with demand. During the following year, a record 20 million copies of Trivial Pursuit were sold in the USA alone. This amazing figure meant that one in every eleven Americans now had a semilegitimate opportunity to belittle friends and neighbours for not knowing what language Brazilians speak.

Selchow and Righter arranged distribution rights for Europe and Australia, while their accountants (conservatively) celebrated retail sales exceeding $1 billion.

What's the secret of success?

Why has Trivial Pursuit been so successful? Part of the answer is that it stood out in stark contrast to the video games of the 1980s. Here was something that enabled you to interact with people rather than a console. It was fun to play Trivial Pursuit with friends while you ate pizza and drank a traditional hops-containing beverage effervesced with carbon dioxide. There's the joy of dredging from the depths of your brain answers to questions that you didn't even realise that you knew. But perhaps the real secret of the game is that while you might not know who composed *The Magic Flute*, correctly naming the title song from *The OC* may propel you to a victory over that know-it-all classical music buff.

And, yes, the second person to set foot on the Moon was Edwin (Buzz) Aldrin. Lobster blood is usually colourless. Brazilians speak Portuguese. Phantom Planet performs *The OC*'s jaunty 'California'.

Early days:
A monopoly on board games

When you play Monopoly your objective is to remain financially solvent while forcing your opponents into pecuniary torment.

Scott Abbott and Chris Haney had many moments in the early days of developing Trivial Pursuit when they faced real-life financial crises. But if you want to devise a board game, Monopoly isn't a bad one to select as your benchmark.

Monopoly's manufacturers, Parker Brothers, explain that the game was invented in the 1930s during the Great Depression. Charles Darrow, an unemployed Pennsylvanian engineer, came up with the idea. However, the game was rejected for being too complicated and taking too long to play.

Undaunted, Charles decided to go it alone, producing 5000 sets. He named the locations on the board after streets and railway stations in Atlantic City. Soon, Monopoly was selling so well that Parker Brothers made Charles an offer he couldn't refuse. Over 200 million Monopoly sets have been sold worldwide, making it the best-selling privately patented board game in history.

You're kidding:
What's in a name?

Plenty, is the answer, according to Fred Worth. Fred is author of *The Trivia Encyclopaedia*, published in 1974. Concerned that others might copy the material in his book, he set a cunning trap – he deliberately included a question with a fake answer. This way, if anyone purloined his material, he would have proof. He realised that facts are part of the public record and cannot be copyrighted. However, he could claim rights, he decided, to his own original 'facts'.

Years later, Fred was reading through the questions in Trivial Pursuit. With a jolt, he realised that a number of the questions looked familiar. Very familiar. Many, he believed, were lifted from his *Trivia Encyclopaedia*. Reading on, he came across his own fabricated question and answer.

Fred's trick question, reproduced in Trivial Pursuit, was, 'What was Columbo's first name?' Columbo was a long-running, popular television series, starring Peter Falk as the humble, dishevelled detective. 'Philip' was the name that Fred had concocted as the answer (it was never given in the show; when Columbo is asked what his first name is, he says 'Lieutenant'). Trivial Pursuit, he decided, had taken the bait. He now had proof that the board game had lifted his material.

In 1984, Fred's lawyers filed a lawsuit against Trivial Pursuit. The suit claimed $300 million in damages for copyright infringement. The suit claimed that, as well as including Fred Worth's worthless fabricated question, the board game had even copied typographical errors and misprints from Fred's book.

In response, Trivial Pursuit said that, yes, they had copied from the book, but they had also used material from many other sources.

The judge dismissed the suit. Four years later, Fred's lawyers took the case to the United States Supreme Court, which rejected the appeal. Thus ended Fred's pursuit of Pursuit.

HOW IT WORKS:
Pursuit of the game

The first edition of Trivial Pursuit came in a distinctive blue box. Colours are key to the game, with each of the six subject areas being distinctively hued. Geography questions are blue. Blue for the ocean? Science and nature, our favourites, naturally, green. Brown equals art and literature, the colour of the mouldering cover of a classic book, perhaps. Entertainment, pink. And we won't speculate further on colour choice.

Play singly or as a team. Roll a dice. You'll have a couple of choices of the colour square – and hence question category – to move to. If you are part of a team, bicker to determine which square to move to. Try to correctly answer the question. Some of the squares permit you a wedge of that colour if you are right. Keep rolling and showing off. Collect a full complement of category wedges, be fortunate with your dice throws and answer a final question from a group that your opponent thinks is your weakness. Gloat.

Today, you can buy Trivial Pursuit in nineteen languages and thirty-three countries. As well as the original and Silver Screen versions, there is the baby boomers' edition ('6 new categories, 6000 new questions from atomic power to flower power'), the all-star sports edition and a young players' edition. There is also the 1990s edition and the book lover's edition. You will be little surprised to learn of the *Lord of the Rings* movie trilogy collector's edition. There is even the Trivial Pursuit pop culture DVD edition.

References

Internet

Trivial Pursuit home page, at <www.trivialpursuit.com/trivialpursuit/about.html>.

Columbo, 'Just one more thing …' website, at <www.columbo-site.freeuk.com/firstnamecourt.htm>.

Founders, at <www.collections.ic.gc.ca/heirloom_series/volume4/216-219.htm>.

The Greatest Discovery Since Fire

Zap! Percy Spencer cooked up an invention that makes waves

'It's the greatest thing since sliced bread.' Such an accolade is often bestowed on an invention these days, but, for some reason, it always seems difficult to surpass the invention of sliced bread. Very few innovations have been heralded as the greatest thing ever, better even than sliced bread. However, one product, which is now found in almost every home in the Western world, was described in the 1960s as 'the greatest discovery since fire'. Now that's an accolade.

It takes more than a chance observation to produce an invention that rivals fire. You also need to throw in an insightful mind to interpret the observation and turn it into a bright idea, a practical engineer to develop the idea into a product and a cunning marketer to popularise it to the world and turn it into a multimillion dollar industry. Nonetheless, it is with an observation that this story begins.

Percy Spencer was working for Raytheon in Massachusetts, the largest electronics manufacturer in the USA. The company made radar equipment for military use and for the Massachusetts Institute of Technology as part of the second-highest priority Second World War military project after the Manhattan Project, which developed the atomic bomb.

Around the end of the war, Percy observed that a chocolate bar in his pocket melted when he stood close to a magnetron, which generates the radio signals at the heart of a radar set. In the 1800s, Louis Pasteur used to say that, when it comes to observations, chance favours the prepared mind (which, of course, he would say in French). Percy's mind was, indeed, prepared for identifying something special.

Microwave moments

Rather than ignore the chance observation of his chocolate-bar mishap, as others had presumably done when engineers had warmed themselves by stacks of magnetrons during the war, Percy sprang into action to see if other foods could be cooked by the magnetron's emitted high-frequency radiowaves – known as microwaves.

He gestured for a colleague to come over to him by giving a small wave – a microwave, so to speak. He asked the colleague to go out and buy a bag of popcorn, which he placed near the magnetron. Amazed, they watched as popcorn exploded all over the lab. So popcorn has its place in history as the first food ever to be deliberately cooked by microwaves.

Not afraid of a messy lab, the next morning Percy tried cooking an egg. He placed an uncooked egg in its shell inside a kettle and moved the magnetron against the kettle's opening. One of his colleagues, sceptical that Percy's experiment could actually work, looked into the kettle just as the egg exploded. The colleague caught a face full of cooked egg, setting himself up for what we imagine was the best joke Percy had ever delivered.

Early days:
Self-made Maine man

It was almost as though Percy's life had prepared him for the moment when the chocolate bar in his pocket melted.

He was born in 1894 in a remote farm community in Maine, in northeast USA. Percy was an unlucky child, for, as Gilbert and Sullivan would say, he was orphaned often: once as a toddler when his father died and his mother left him to his uncle, then a second time aged 7, when his uncle died. Aged 12 Percy quit his country chores and school for work, and by 16 he had a job installing the fairly new technology of electricity into a local paper mill. As a competent but self-taught electrician, he was inspired by the heroic wireless operators on the sinking *Titanic* in 1912 to join the US Navy's radio school.

After leaving the Navy, Percy worked for the Wireless Specialty Apparatus Company in Boston, being promoted to manage wireless equipment production during the Second World War. In the 1920s and 1930s, as he worked his way up through the Raytheon company, the self-educated experimentalist with no more than a primary school education would work with Massachusetts Institute of Technology's top scientists. At Raytheon during the Second World War he developed methods that enabled his team to produce 2600 magnetrons a day; at the start of the war, it had taken specialists a week to make a single one. He also improved the magnetron's efficiency and effectiveness on fighter planes in combat. All this experience was preparing Percy's mind for the idea that came to him after the war when the magnetron's high-frequency radiowaves melted his chocolate bar.

Percy applied in 1945 for the first patent for a microwave oven, which he envisaged would cook food as it moved on a conveyor belt through magnetron waves. But cooking wasn't the only use he saw for microwave ovens. He imagined it would one day be used for a wide range of applications, from ink drying to tobacco curing.

The first microwave oven was rubbish

With the backing of Laurence Marshall, head of Raytheon, Percy and his colleagues began building microwave ovens. A prototype was no more than a radar tube sticking out the back of a rubbish bin lying on its side. The engineers, who were used to making specialist equipment for trained military users, had to come up with something foolproof and easy to use. Percy employed a young engineer, Marvin Bock, to develop a microwave oven for sale to the public.

Marvin had to balance the competing priorities of ideal frequency, oven size and food orientation to ensure balanced cooking. A wavelength of 12.24 centimetres

(a frequency of 2450 megahertz) worked well – it didn't interfere with radio transmissions, and it was small enough to ensure food cooked evenly. After many experiments to perfect the magnetron tube, the power supply and the rotation of food through the microwaves, Marvin was ready to subject food to his product.

His notebooks record his culinary exploits. Potatoes cooked in a minute – 'the flavour was good but the potato was not crisp.' Brussels sprouts cooked for 1 minute 15 seconds – 'the flavour was dry and not good.' For meat, he commented that 'Steak doesn't brown.' Marvin does not note whether there were others at his horrible dinner party and, if so, whether they ever came back.

In 1947 Raytheon produced the first commercial microwave oven, named the Radarange following an employee competition. The monster microwave was just under 2 metres high, almost 1 metre deep and wide, weighed 340 kilograms, threw out three times the amount of microwave energy produced by today's microwave ovens and required a system of water pipes to keep it cool. It was a bargain at today's equivalent of $40 000. In short, it was not something that was going to catch on quickly in a domestic kitchen.

You're kidding:
Dangers and myths

Microwaves cannot escape the oven's Faraday cage enclosure. The metal mesh on the door allows short wavelength visible light through, but longer wavelength microwaves cannot escape. The microwaves shut off when the timer ends or the door opens. So, you do not need to worry about being nuked by a microwave.

Due to uneven cooking that does not kill contaminating bacteria, there is a risk, when reheating previously-cooked food, of food poisoning. Another danger is distilled water, which can heat to above boiling temperature without actually boiling, and can then boil explosively when stirred or disturbed. Also, buildup of steam causes a risk of explosion when microwaving closed containers or, in a nod to Percy's original cooking experiment, eggs.

Metal, such as aluminium foil, blocks and reflects microwaves, so can cause sparks. But it won't cause your microwave to catch fire or explode.

There are claims that microwave cooking destroys the health benefits of food or turns nutrients into carcinogens, but none of these claims has been supported by proper scientific research. In actual fact, some vitamins are retained in microwaved vegetables much better than in boiled vegetables.

There are stories about telecommunications workers and hotel chefs being cooked by microwaves, and even one about an old woman accidentally killing a poodle by trying to dry her pet in a microwave oven. Such stories are urban myths, and have been proven to simply not be true.

Marketing genius

Enter Raytheon supersalesman Art Welch, who marketed the microwave ovens to restaurants, train buffets and ships' kitchens. Art, wearing an ostentatious diamond ring and a tie with a duck on it, conducted a spectacular stage show to highlight the cooking speed: the show involved chefs, microwave ovens and a giant clock. The advertising campaign across the USA was a success, but by 1948 pretty much everyone who needed a $40 000 microwave had one.

Raytheon launched an updated Radarange in 1953, selling 10 000 of them by 1967. The first home microwave oven was on sale in 1955, but at half the cost of a Radarange it was still

not cheap enough to make an impact. The technology developed rapidly, initially by defence contractors who were experienced with the magnetron, thus bringing costs down. Through its retail arm, Amana, Raytheon launched a sleek, elegant microwave oven onto the market in 1967, showing it off using a team of demonstrators on a train tour through the USA. The time was right – many households now had two working parents, and ready-made meals or reheating had become the way to make dinner.

By the late 1970s, prices were falling sufficiently to bring microwave ovens within reach of everyday kitchens. By the 1980s, they had morphed from expensive curiosity to cheap kitchen necessity in a hectic world. Microwave ovens are now estimated to be in 95 per cent of American kitchens, and there are more than 200 million microwave ovens in use around the world today.

Amana's advertising campaign in the late 1960s blurted the headline, 'the greatest discovery since fire' across newspapers and magazines. And perhaps that is what the microwave oven for the kitchen was, one of the biggest changes to the way we cook our food since the days of fire. But nobody has yet come up with a microwave fireplace for the living room. And why would they? Who, after all, would want to curl up and have a nice, quiet evening in front of a fire if it were to last only a few minutes?

Percy died in 1970 aged 76. He had lived to see the acceptance of his invention, risen to senior vice-president at Raytheon and had around 150 patents to his name.

HOW IT WORKS:
Turn up the radio

People have used radiation to heat and cook for millennia – sunlight emits radiation at visible (and other) wavelengths; our ancestors used the visible and infrared radiation from fires to cook and stay warm; and electric ovens cook using radiation from an element rather than a gas flame. Radiative heat cooks food from the outside, penetrating food through the process of conduction.

Microwaves, radiowaves with much longer wavelengths, penetrate food and set water, sugar and other molecules in motion. Molecular motion is what creates heat, so this considerably reduces the cooking time. As Raytheon microwave oven demonstrator JoAnne Anderson used to explain, 'there is energy, that is radiowave energy, that basically attacks water molecules in your food . . . it activates those molecules. That causes friction, and friction, we know, is heat. Therefore, your food is cooking itself.'

Thank you, JoAnne.

Many materials, such as glass, paper and foam containers, do not absorb microwaves, so do not heat up in a microwave oven. Heat is conducted to the container by hot food. Waste heat from other elements of the microwave, such as the cooling fan, turntable and light, are discharged through vents in the oven.

References

Internet

American heritage of invention and technology, at <www.inventionandtechnology.com>.
<www.inventionandtechnology.com/xml/2005/4/it_2005_4_feat_4.xml>.
.
<www. nationmaster.com/encyclopedia/Microwave-oven>.
Urban Legends reference pages, at <www.snopes.com>.
Encyclopaedia Britannica online, at <www.britannica.com>.

Texts

Reader's Digest, August 1958.

The Power of Poly-Para-Phenylene Terephthalamide

Stephanie Kwolek weaves a bulletproof idea

Whether it's working in a plant, at a construction site or in a forest. Whether it's walking a beat, escorting a prisoner to his cell or completing a tour of duty. Whether it's winning the America's Cup, the Tour de France or the Boston Marathon. Whether it's protection from the hazards they face on the job every day or the comfort and lightweight strength they rely on to help them achieve their personal best, men and women around the world rely on ...'

This is how DuPont's website, 'The Miracles of Science', spruiks the product that made Stephanie Kwolek her millions. Apart from the line about 'comfort and lightweight strength' and the reference to 'escorting a prisoner to his cell' this piece of advertising prose dreamt up by a company PR hack could be describing a sports drink. Or maybe the invisible woman.

Instead, it is trumpeting one of the world's strongest fibres, with five times the strength of the same mass of steel. The fibre's name is poly-para-phenylene terephthalamide, known to all the world, including prison guards, as Kevlar. The material has numerous applications in addition to the famous bullet-proof vests. It is used in radial tyres and brake pads, fibre optic cables, racing sails, space vehicles, boats, parachutes, skis, and building materials. You will also find the substance in safety helmets and in a host of outdoor sporting gear. Thin, lightweight Kevlar ropes secure massive ships to wharves and secured the airbags that cushioned the landing of the Mars *Pathfinder*.

Worldwide, Kevlar is worth hundreds of millions of dollars per year to DuPont. Learn more from the DuPont website in which our diligent PR employee, clearly on a roll, enthuses about the super fibre helping 'men and women around the world... realize the Power of Performance'. Not just the power of performance, mind you, but the strengthened, specially capitalised version thereof.

Picking up the thread

Born in New Kensington, Pennsylvania, in northeastern USA in 1923, the inventor of Kevlar, Stephanie Kwolek, grew up with a strong interest in science and medicine.

The Smithsonian Institute reports that, as a child, Stephanie loved exploring the natural world with her father, collecting wildflowers and seeds for her scrapbook. She also had an early interest in technology. 'When I was six,' she recalled, 'I loved to use my mother's sewing machine. I was forbidden to use the sewing machine, but I would sneak in when my mother went shopping.'

Stephanie's father, a foundry worker, died when Stephanie was just 10 years old. Lacking the funds needed to become a doctor, Stephanie completed a

Early days:
Explosive growth

Éleuthère Irénée du Pont, you will not be surprised to learn, was a Frenchman. His father, Pierre-Samuel, was an econ-omist, a blueblood who didn't think much of the policies of the radical republicans in late eighteenth century France. In fact, Pierre-Samuel was one of the initiators of the French Revolution through his advocacy of the Tennis Court oath. We've been responsible ourselves for one or two of these, but few resulting in a wholesale guillotining regimen, even though we could nominate a couple of deserving candidates.

The evocatively dubbed oath is named after an historic act of independence by the French working class in 1789. Locked out of their regular Versailles meeting hall, the worker rep-resentatives adjourned to a nearby tennis court where they committed to work together for a written constitution for the country.

Éleuthère Irénée worked at the French royal powder works, a business that must have been doing a roaring trade in those tumultuous times. In 1800, the 30-year-old set sail for the USA. Discovering that the Americans liked peace as little as his compatriots, Éleuthère Irénée soon established a gunpowder plant near Wilmington, Delaware. Business boomed. The War of 1812 was great for Éleuthère Irénée. Not only was he selling wagonloads of the black powder, but his adopted USA was also raising arms against the great foe of the French, the British.

In 1833, Éleuthère Irénée named his business E. I. du Pont de Nemours and Co. Thus began one of the world's behemoth manufacturers of chemicals, plastics and synthetic fibres.

chemistry degree at the Carnegie Institute of Technology (now Carnegie–Mellon University) in nearby Pittsburgh. She was then successful in obtaining a research position in 1946 with DuPont's textile fibres laboratory located at Buffalo in New York state.

DuPont was the perfect environment for the highly inventive Stephanie. One of the world's largest businesses, the company had been founded as a manufacturer of gunpowder in the early 1800s by the French-born Eleuthère Irénée du Pont de Nemours, one of the world's great science entrepreneurs.

Stephanie originally intended to work for DuPont for only a few years – just long enough to save enough money for medical school, but she enjoyed her work so much that she forgot about medical school.

Stephanie thrived in a work environment that offered a great deal of freedom. DuPont also offered a tremendous amount of excitement that was associated with the numerous inventions that its researchers were coming up with.

However, there were challenges along the way. When Stephanie joined DuPont in 1946, women in science were regarded as an oddity. Those who got laboratory jobs would stay only a few years before being encouraged to move into 'women's fields'. Female researchers were not promoted as rapidly as men and many lasted only several years before opting for the well-trodden path to teaching, often at girls' schools. But Stephanie stuck it out.

Perfect alignment

The Second World War had just ended when Stephanie began at DuPont. Her early work was with polymers – giant molecules comprised of many linked, repeated chemical units. (The single building block of a polymer is a monomer, so polymers consist of numerous 'mers'.) Most natural organic materials are polymers – including proteins, wood, chitin (the hard shells that protect bugs), rubber, latex, resins and silk. It was a fine time for a young chemist to be joining this field. The war had caused shortages of many of these natural polymeric materials so there was a massive market for laboratory-synthesised replacements. Think nylon, polyesters and various types of synthetic rubber. Some of the new plastics were perfect replacements for metal in machinery, safety helmets and equipment that had to endure high temperatures.

In 1950, Stephanie moved to DuPont's new research laboratory in Wilmington, Delaware, where she sought to discover polymers that could be spun into fibres at room temperature, thus reducing the expense and inconvenience of heating. In the laboratory she successfully produced a range of strong, rigid fibres from petroleum.

Her breakthrough came when she observed new polymers whose molecules lined up in a highly regular, parallel, rodlike arrangement in solution. No one had ever seen such a thing. Stephanie and her team had prepared what came to be known as liquid crystal polymers.

Most polymer solutions are thick and syrupy, but the solution that Stephanie created had an unusually low viscosity. This one flowed like water and looked like milk.

The next step for Stephanie was to convince her technician to put the hazy solution through the spinneret, which produced fibres. The technician believed that the milky colour was caused by solid particles that would gum up the spinneret's tiny holes. Eventually, he gave it a whirl. The result – fibres with unprecedented stiffness and strength. DuPont named the product Kevlar, releasing it commercially in 1971.

Stephanie Kwolek would ultimately obtain twenty-eight patents during her 40-year career as a research scientist at DuPont. She has won innumerable awards for her work from science and technology organisations. Stephanie is particularly proud of the fact that her invention of Kevlar has saved the lives of well over 2000 police officers.

You're kidding:
Unholy priest

Kevlar might be best known for its application in bulletproof vests, but it is far from the first material to be used in this way. Picture Ned Kelly in that famous home-made suit of armour.

In the late 1800s, a number of enterprising individuals toured the USA and Europe, demonstrating what they claimed were bullet-stopping clothes. One, an Austrian tailor by the name of Herr Dowe, even allowed himself to be shot at. 'I feel nothing,' he announced. Unfortunately for Herr Dowe, an army officer who eventually managed to inspect the tailor's clothing did not feel nothing. He felt plenty – hidden beneath the outer material were wire netting and a solid substance something like cement. This bulky, heavy material was far from ideal for wearing in combat or in dangerous situations.

Enter Casimir Zeglen, a young Chicago priest, who had recently arrived from Poland. Horrified by the 1893 fatal shooting of the mayor of Chicago, Carter Harrison, Brother Casimir set out to perfect clothing that could stop a bullet. After experimenting with innumerable materials, including, oddly enough, moss, Casimir settled on another natural product – silk. He wasn't sure how best to weave it, a problem solved by a visit to famous weaving mills in Austria and Germany, where he found that very tightly woven silk sandwiching a 1.6-millimetre-thick steel plate was as effective as a suit of armour in deflecting bullets and somewhat less conspicuous.

It would seem that there is only one way to prove the effectiveness of a bulletproof vest. By good fortune, Count Zarnecki, an Austrian army officer and close friend of Casimir, was visiting the USA. In a highly publicised event, Casimir, resplendent in his newsilk and steel vest, appeared on stage at a Manhattan theatre with the count. After a few warmup shots at inanimate objects,

You're kidding cont.

the count turned towards Casimir. From a distance of eight paces, he raised his Colt revolver, carefully aimed it at the middle of his friend's chest and fired. Thankfully, Casimir survived.

Despite putting himself through a death-defying ordeal, Casimir did not make a fortune from his invention. The steel made the clothing too heavy, the woven silk made it too expensive. In any case, the priesthood was a difficult platform from which to launch a new business.

HOW IT WORKS:
Vested interests

The trigger is squeezed. The firing pin strikes the cartridge case. The powder in the case explodes, forming a rapidly expanding gas. The gas forces the bullet at high speed along the bore. The bullet exits the bore at a speed of 1000 metres per second. This might be slower than Superman, but it is more than twice the speed of sound.

A fraction of a second later, the bullet strikes you. Another fraction of a second later you hear the explosion from the firing gun. A second or two later you bless Stephanie Kwolek and her invention of the Kevlar in your bulletproof vest, which has saved your life.

Here's how the vest works. Many layers of heavy weave Kevlar cloth are stitched together like a quilt. Kevlar excels here because its fibres are so strong. It takes incredible amounts of energy to stretch them. It is easier to stretch steel than to stretch Kevlar. The cloth in your vest consists of massively strong, densely intertwined Kevlar fibres. As the bullet hits the outermost layers, it is flattened and deformed. Much of the energy is consumed by the fibres around the point of impact, which stretch ever so slightly. The process spreads the bullet's energy through the material so that only a fraction reaches your body. It will still hurt and leave a bruise, much as if you are struck by a cricket ball, but you will have another innings.

References

Internet

Bulletproof vest, at <www.ideafinder.com/history/inventions/story082.htm>.
<www. whyfiles.larc.nasa.gov/text/kids/Media_Zone/images/kwolek 01.html>.
Women Inventors, at <www.hrw.com/science/si-science/chemistry/careers/innovative _lives/womeninventors.html>.
DuPont website, at <www.dupont.com>.

Texts

Brooklyn Eagle, 9 October 1902.
Encarta Encyclopaedia, 1997.
Collins, P. 2005, 'Histories: The bulletproof priest', *New Scientist*, issue 2496, 23 April.

Mightier Than a Sword

How Paul Fisher wrote his name in history

During the space race of the 1960s, American astronauts were faced with the problem of how to write in the absence of the gravity required to draw ink from a pen. To solve the problem, NASA spent more than $1 million to come up with a pen that could write in the vacuum of space. The Soviets faced the same dilemma. They also came up with a solution – the cosmonauts used pencils.

The story illustrates perfectly the benefits of using a simple, cheap solution over an expensive, high-tech one. Unfortunately, though, the story is not true. Astronauts and cosmonauts both used pencils in the early days of space flight, but the graphite tips would break off and become a hazard as they floated around the spacecraft. The solution to this problem was presented to NASA by inventor Paul Fisher, whose Space Pen has subsequently been used on all American and Russian space flights. You could say that Paul had the Write Stuff.

To fully appreciate the true story of the Space Pen, we must visit its history. And it is lucky that the pen was invented, for without it, how could its history have been written?

Writing through the ages

The pen's ancestor was the Chinese writing brush, used as long ago as 3000 years; the Egyptians used pen-like reeds around 2300 years ago. By the sixth century, the Romans were using quills – bird feathers whose hollow tubes could store ink – which were the favoured writing implement for more than a thousand years. In fact, the word 'pen' comes from *penna*, the Latin word for feather.

The first practical fountain pen was made by the American inventor Lewis Waterman in 1884, although similar pens had appeared throughout history – including a bronze pen found in the ruins of Pompeii. More than a thousand patents were registered for fountain pen improvements, and business boomed for the next 60 years. Then, with the appearance of an invention by the Biro brothers, fountain pen businesses dried up faster than fountain pen ink.

In 1945, Chicago businessman Milton Reynolds saw a biro in a Buenos Aires shop and immediately recognised the potential for sales in the USA. He bought several samples, developed his own version and launched the Reynolds Rocket from a New York City department store. Milton sold $100 000 worth on the first day, and within three months had sold 2 million pens. With a profit of almost $2 million, here was one of the most successful product launches in American history.

However, the pens were notorious for leaking, which had a number of consequences. First, it introduced the fashion mistake known as the pocket protector. Second, it led to people returning their pens, which in turn led Milton to ask a young inventor named Paul Fisher to see if he could make a better pen.

The rest, as they almost say, is his story.

The pen men

Paul Fisher was born in 1914, the son of a Methodist minister. When he was growing up in Kansas he was a curious and inventive child, who made a radio out of a cereal box, wires and a crystal. After completing

Early days:
Basic background to the biro

Hungarian Laszlo Biro had studied disciplines as varied as medicine, art and hypnotism. But it was as a newspaper editor in 1935 that he noticed that newsprint dried quickly. He tried using newspaper ink in a fountain pen but found that its thickness clogged the nib. So Laszlo and his brother Georg, a chemist, invented a new tip made of a tiny metal ball that turned freely in a socket, picking up ink from a tube and depositing it on paper. Although John Lourd had patented a rollerball pen in 1888 and designs by Galileo in the 1600s depicted a ballpoint pen, the Biro brothers obtained a patent for their invention in 1938.

After escaping to Argentina during the Second World War, the Biros developed an improved ballpoint. Already a success in Argentina, the new pens were picked up by the British government. The innovation enabled RAF pilots to do navigational work at high altitudes, where the low pressure would normally cause fountain pens to leak (perhaps they, too, had not thought to use pencils). Soon the Biro brothers joined Hoover, Xerox and Esky in having their trade name enter common usage in English and other languages.

In France, Marcel Bich obtained the rights from the Biros to improve their design. He dropped the 'h' from his name to launch the Bic and be immortalised in French, among other languages. Today, 14 million Bics are sold each day.

Just as the biro has entered the English language, the Space Pen has entered our culture. An entire *Seinfeld* episode centred on a Space Pen. A misunderstanding between Jerry and a friend of his father leads to Jerry accepting a Space Pen as a gift, and then returning it college in Kansas and Iowa, he had various jobs, including as a bread store manager, a truck driver, an accountant and, in a position that gave him the background he needed for inventing a better ballpoint pen, the manager of a ballbearing company.

Paul worked for Milton Reynolds in 1945 but gave up the chance to join Milton's soon to be million-dollar business. Milton gave Paul a biro and asked him to perfect it. After investigating it for a couple of days, Paul explained to his boss that the pen's basic principle was no good and walked out.

However, Paul stayed involved in the pen business, founding the Fisher Pen Company in 1948. He made his first million dollars in 1953, with the invention of the Universal Refill. This was a cartridge that could fit most of the dozens of ballpoints on the market in the 1950s, which, until then, had required their own unique refills.

Paul's company continued to improve the ink refills until he realised in the midst of the space race how to really leave an indelible mark on the world. It occurred to him that astronauts would need a sealed and pressurised pen that would work in the vacuum of space.

Space-age vision

Paul modestly says the key to his success is luck rather than smarts. But it must also involve plenty of rapid eye movement, because he says that the solution to leaky pens came to him in a dream. In his dream, his recently-deceased father told him to add *rosin*, a plant-based liquid used in pens to assist ink flow. Back in the waking world, the company's chemist laughed at Paul, but humoured him by experimenting with adding different types and amounts of rosin. Finally, it occurred to the chemist that Paul had meant the much more viscous *resin*. He added 2 per cent resin to the ink and found that, even at high pressure, the cartridge wouldn't leak. This allowed the so-called thixotropic ink to be pushed out of a cartridge by the pressure of a small volume of gas, so gravity was no longer necessary.

Further developments

In 1965, Paul took out a patent on the anti-gravity pen. As well as writing in a vacuum and without the assistance of gravity, the pen could write upside down, over oil and grease, on plastic, at temperatures below freezing or above boiling and under water. He sent samples to Houston Space Center where, after being tested thoroughly, the pens were adopted for use in space.

Today, the Fisher Space Pen Company has annual revenues of $8 million and employs seventy people. Paul is still inventing pens, having recently launched the Millennium pen, which is guaranteed to write for a thousand years and so outlast the period over which the quill dominated writing. But Paul hasn't been a success in everything: in 1960, he unsuccessfully ran in the 1960 New Hampshire presidential primary elections . . . against John F. Kennedy and Richard Nixon.

Paul spent more than a million dollars perfecting his Space Pen. He says he has made more than

HOW IT WORKS:
A ball is the point

The metal ball on a ballpoint is like the ball on a stick of roll-on deodorant; it acts as a cap to prevent the liquid drying out, and to control the amount of liquid that flows out. Ink is made of a dye (from pigments such as carbon for black or iron for blue) dissolved in a solvent (such as linseed, rosin or kerosene) and a stabilising agent. It is thick enough not to leak, but thin enough to be drawn out by gravity. A small hole in the pen casing prevents the pressure from dropping as the ink is drawn out (if this hole is blocked, the ink cannot flow).

A Space Pen's ink cartridge is pressurised and the ink is like a thick rubber cement. The very hard tungsten carbide ball shears across the ink to liquefy it, while the 340 kPa of pressure from nitrogen gas in the cartridge constantly forces the ink out the nib. A plug separates the ink from the gas, sliding down as the ink is used up.

The marketing blurb says the Fisher Space Pen will last for 100 years on Earth and in space, but Paul Fisher does admit one fault – sometimes it spells incorrectly.

10 000 pens that didn't work, using a simple scientific technique known as trial and error. However, the funding for his research did not come from NASA, and the story of the national space agency spending millions of dollars on research when a simple pencil would suffice is simply a myth. In December 1967, NASA purchased 400 Fisher Space Pens for $2.95 each – a total of $1180, well short of a million.

SUCCESS SCALE

MARKETING

Space Pens have the benefit of a great story (saving *Apollo* 11 astronauts; see page 94). The lucky timing of Paul's space pen during the space race must have helped popularise the pen.

INNOVATION

Coming up with a solution in a dream is a handy way of avoiding expensive research and development. Paul made a good product better, and continues to invent better pens.

FINANCE

The invention of better writing implements has been rewarding to the Biros, Bics and Fishers.

References

Internet

How Stuff Works, at <www.home.howstuffworks.com>.
Space Pen, at <www.spacepen.com>.
Urban Legends reference pages, at <www.snopes.com>.
Wikipedia, at <www.en.wikipedia.org>.
<www.ideafinder.com>.
<www.inventors.about.com>.
Encyclopaedia Britannica online, at <www.britannica.com>.

Liquid Assets

Bette Nesmith made no mistake when she helped millions of typists correct theirs

Most inventions happen in one fell swoop. Some bright person sees a need or market for a new product and carefully and painstakingly produces a prototype. The device is patented, the production lines are powered up and, if the inventor is particularly fortunate, they watch the zeroes rapidly being added to their bank balance.

As a way of neatly presenting words on paper, the typewriter was arguably the most significant business and communica-

tion tool until the birth of personal computers in the 1980s. No one person can claim credit for inventing the typewriter. This was a tool that came into being through a series of shambolic, stuttering steps. Early typewriters, in the 1800s, were the size of pianos, with intricate arrangements of keys, levers and rods.

It was in 1867 that an American inventor, Christopher Sholes, built a practical machine that served as the basis for the next hundred years of typewriters. There was a keyboard that was pretty much identical to the computer keyboard used today. A writer employing the Sholes machine and its successors would attach a sheet of paper to a cylinder that moved from left to right. Push down the 'A' button and a mechanism would cause a metal block with an embossed 'A' to pound on a ribbon impregnated with black ink. The ink would mark the letter on the page, the cylinder would progress fractionally to the right and all was set for the next letter. To add to the excitement, later machines incorporated sound effects in the form of a little bell that would alert the typist that they were approaching the right-hand edge of the page. At this point judicious mental calulations were needed to establish if there was sufficient space for one more word. If not, the typist would perform a 'carriage return', using a lever to push the cylinder with attached page to its leftmost position and up one line to start a new line of text.

Now we leap forward many decades to the life and times of the true star of this story, a woman who made first a living, and second, a fortune from the typewriter.

Creating the perfect typist

Bette McMurray was born in Dallas, Texas, in 1924. She grew up in San Antonio to the south, attending Alamo Heights School. Bette left school at 17 and found a job as a secretary for a law firm. Her typing was far from perfect. Fortunately, the company sent her to night school to undertake secretarial training.

In 1942, Bette married a soldier, Warren. A year later, she gave birth to her only child, Michael. The marriage didn't work out, and Bette and Warren divorced in 1946. Now Bette was the sole provider for herself and her son.

The fact that you are reading here about Bette tells you that she is destined for greatness. By a quirk of history, she is not the only person in this yarn – or her family – who made good. Ah, you might say, the son of a millionaire is not going to find himself short of a few dollars. But here's the thing – Michael was to become highly successful in a field very different from his mother's.

Things were far from good for Bette in the late 1940s. With a limited income, it was a struggle to make ends meet. Her love of

Early days:
The perfect letter

In the days before Liquid Paper, and decades before the computer 'delete' and 'backspace' buttons and spellcheckers, a business letter was supposed to have no mistakes and no visible corrections. This meant that, in addition to being fast, the typist had to be highly accurate.

Perfection, as they say, exists only in a job application. Typists used a variety of techniques for correcting the odd typo. Most common was a hard, stiff rubber eraser. The typewriter eraser was usually the shape of a thin, flat disk, allowing the dextrous typist to rub away a mistyped letter.

Paper companies were in on this sleight of hand. They produced thick typing paper that could withstand rubbing. There was even erasable bond typewriter paper that incorporated a thin layer of material that prevented ink from penetrating the paper. The layer with the offending typo could easily be scratched away.

Any significant errors or a change of mind by the boss would mean that the entire letter had to be retyped.

art and her need of additional income saw Bette working a second job as a freelance artist.

A few years later, the 27 year old landed a job as a secretary at the Texas Bank and Trust in Dallas. The bank had new electric typewriters, which made typing easier and faster. (An electric motor controlled keystrokes, so the typist needed to use less force to strike the keys and the letters printed more uniformly, thanks to the constant pressure with which typewriter ribbon was struck. The motor also returned the cylinder and attended to line spacing.) But electric typewriters had one disadvantage over their manual counterparts: their carbon-film ribbons made corrections far more difficult. Trying to rub out mistakes left unsightly black smudges on the page.

Whitewash

Bette wanted her typing to appear perfect. From her art background, she knew that painters could brush over flaws and smudges on canvas, rendering them invisible. Perhaps the same approach could work for typewritten mistakes.

At home, she mixed various hues of water-based paint until she had a good match for the

Texas Bank and Trust stationery. She poured the paint into a little bottle, packing it into her bag, along with a little paint brush, the next morning.

Bette was delighted with how well her invention hid her typing errors. She dubbed it Mistake Out. Before long, other secretaries were borrowing the paint and using it themselves.

Keen to improve the new product, Bette checked the library for paint formulae and sought advice from a chemistry teacher at St Mark's, a private boys' school in Dallas. A paint company employee showed her how to grind and mix paint. A local office supply dealer provided suggestions. Soon Bette was hard at work in her kitchen, blending batches of Mistake Out in an old electric food mixer.

The production line expanded into the garage. Deciding that Liquid Paper was a better name than Mistake Out, Bette obtained a patent and trademark.

Her son, Michael, and his friends helped with bottling. In an effort to boost sales, Bette contacted IBM, asking them to market her invention. They declined. (This was an IBM blunder on par with their Chairman Thomas Watson's 1943 assertion that 'I think there is a world market for maybe five

You're kidding:
One, two, three . . .

According to the *Guinness Book of World Records*, Barbara Blackburn of Salem, Oregon, is the world's fastest typist. She can maintain a bewitchingly fast 150 words per minute for fifty minutes. That represents some 37 500 keystrokes. If she could keep up that pace, she'd be able to type this entire book in less than six hours.

Her top speed was 212 words per minute. Her secret – the Dvorak keyboard – and great typing skills. This distinctive keyboard has vowels on one side and consonants on the other, with the most frequently used letters on the centre row.

If it's perseverance that impresses you, you'll enjoy learning about Les Stewart from Mudjimba Beach, Queensland. In 1998, after sixteen years at the keyboard, Les achieved his ambition of typing all numbers from one to one million in words on his manual typewriter. The feat took seven typewriters, 1000 ink ribbons and 19 990 sheets of paper.

computers.' In 1968, an engineer at the Advanced Computing Systems Division of the company, commenting on the microchip, would ask: 'But what is it good for?')

It was a small item on Liquid Paper in a 1958 office trade magazine that provided the necessary sales boost. Five hundred orders arrived from across the USA. Then, General Electric placed an order for over 400 bottles in three colours. This request repre- sented four times Bette's monthly production.

Earning from their mistakes

Bette was profiting from the mistakes of typists across the country. But she was also to profit, albeit inadvertently, from a mistake that she made herself. She accidentally added her company's name to a letter typed for her employer. The company sacked her. Now she could spend more time in the kitchen – producing Liquid Paper.

In 1962, Bette married Robert Graham. Combining her first husband's surname with her new husband's name, she became Bette Nesmith Graham. Together they travelled across the USA marketing Liquid Paper. The road trip helped boost demand. This was very necessary because although Bette was working nights and weekends in order to meet demand, her company was making little money. In fact, for quite some time expenses exceeded income.

Thankfully, typists America-wide were making sufficient mistakes to push Bette's white correcting liquid into the black. By 1964 demand for Liquid Paper required her to take on more staff and to increase weekly production tenfold, from 500 to 5000 bottles.

In 1968 the company sold a million bottles and moved into its own plant. Bette had been producing Liquid Paper from her North Dallas home for seventeen years. By now, there were nineteen employees working in a totally automated factory. Demand was such that, in the

mid 1970s, the company moved again into an even larger building in Dallas. Now the plant could turn out 500 bottles every minute. In the following year, 1976, the Liquid Paper Corporation produced 25 million bottles. Two hundred employees distributed the typists' little secret to more than thirty countries.

Bette had a strong social consciousness. She also encouraged her employees to participate in business decision making. She wanted people to be happy at work and so the plant included a library, a childcare centre and gardens.

In 1979, Bette sold the company – which now had sales in excess of $38 million – to the Gillette Corporation for $47.5 million and royalties. Sadly, just six months after the sale, she unexpectedly died. She was only 56 years old. Bette left part of her fortune to women's welfare foundations and to support for the arts.

Those who worked with her speak of the belief that Bette had in herself. She was a spiritual woman with ambitions that were more than just capitalistic. Bette strongly believed that if you do things for the right reason, in the right ways, the success will come.

By now, baby boomers will have worked out how Bette's son

made a fortune independently of his mother. Michael, who spent his formative years pouring liquid paper into bottles for the fledgling business, was a member of the incredibly successful 1960s made-for-television pop group, the Monkees.

HOW IT WORKS:
qwerty keyboard

As typists learnt to type more swiftly, a problem emerged with Christopher Sholes's keyboard. Type bars linking the keys to embossed letters didn't have sufficient time to fall back into place before the next key was struck. If two bars, say, the 'r' and the 't', were next to each other, they would sometimes jam. The solution, decided Christopher, was to rearrange the keyboard, keeping common pairs of letters well away from each other.

Hence the so-called qwerty keyboard, that odd word representing the first six letters on the left side of the top row. It is fitting that the longest word you can type using the top row of letters on a standard typewriter keyboard is 'typewriter'.

Christopher did a fine job. The most common words of two or more letters in the English language are, in order: 'the', 'be', 'of', 'and', 'in', 'he', 'to' and 'have'. None of these words has letters that adjoin on the keyboard.

So the layout today of virtually every English-language computer keyboard is based on a nineteenth century workaround for jamming keys. Even our shift key harks back to early machines. In order to type capital letters, each typing bar included two embossed letters – one lower case and the other its capital form. By holding down the shift key, the typist shifted up the cylinder so that it was the capital letter that struck the printing ribbon.

References

Internet

Bette Nesmith Graham, at <www.gihon.com/pages/bette_graham/bette_graham.html>.
Guinness Book of World Records, at <www.guinnessworldrecords.com>.
PrivateClub magazine, at <www.privateclubs.com/archives/2001-july-aug/life_interview-patricia-hill.htm>.
Wikpedia, at <www.en.wikipedia.org/wiki/Typewriter>.

Texts

Encyclopaedia Britannica online, 2005, at <www.britannica.com>.
Francis, W. N. & H. Kucera 1982, *Frequency Analysis of English Usage*, Houghton Mifflin, Boston.

The Self-Moving Men

The car, and then the production line, steered transport in a new direction

Bridget Driscoll occupies an unhappy chapter in the book of world firsts. One day the fortysomething English woman was walking with her teenage daughter across the grounds of the Crystal Palace in London. Nearby, Arthur Edsall was demonstrating a new machine, which belonged to the Anglo-French Motor Car Company. In an unprecedented accident, Bridget was struck by Arthur driving the recently-invented automobile at what was described by witnesses as the tremendous speed of about 6 kilometres per hour. On 17 August 1896, Bridget became the world's first road fatality.

The coroner investigating the accident claimed that such a terrible situation must never be allowed to happen again. But it did, many times over. Three years later, in New York City, 68-year-old Henry Bliss was alighting from a tram when an electric automobile struck him in what was America's first fatal car accident. Today, around the world, about 1 million people die annually on the road and another 30 million are injured.

The invention of the car gave us fast, comfortable travel but an unfortunate byproduct is a huge number of road accidents. But if you are looking for someone to blame for the introduction of such a deadly machine, the picture is fairly complicated. There are more than 100 000 patents relating to the modern car, the invention of which cannot be attributed to one person.

The German connection

Nicolaus Otto was a German travelling salesman, selling tea, coffee and sugar. Inspired by the development of internal combustion engines in the 1800s – and perhaps by the distances he had to lug bags of coffee – he invented the first practical, petrol-driven, internal combustion engine in 1876. Although he installed it into a two-wheeled vehicle, making the first motorcycle, his engine was the forerunner to future car engines.

In 1885, mechanical engineer Karl Benz invented the first automobile and drove around the streets of Mannheim, Germany.

The patent for his two-seated three-wheeler with a petrol-driven internal combustion engine is considered to be the birth certificate of the car. In 1891, he built a four-wheeled car, but in the quickly developing automotive industry, he wasn't the first to do so.

Gottlieb Daimler, who worked at the company owned by Nicolaus Otto, improved the modern petrol engine in 1885. In 1886 he installed it in a carriage, making the first four-wheeled car. However, Gottlieb didn't like to drive and may never have been behind the wheel of a moving car in his life.

It took a number of years before Daimler and Benz developed from inventors of unique vehicles into car manufacturers. In 1893, Karl developed the Benz Velo, the world's first mass-produced car. But by that stage, the world's first car companies had already formed in France: Panhard and Levassor in 1889, and Peugeot in 1891.

Henry Ford

Charles and Frank Duryea made America's first commercial petrol-powered car in Massachusetts in 1893; Ransom Olds introduced the production line in 1901, producing more than four hundred Oldsmobiles in its first year. Henry Ford invented neither the car nor the production line, but he is known for bringing cars to the people.

Henry was born in Dearborn, Michigan, in 1863, a descendant of Irish and Dutch immigrants. Working on his father's farm he would invent machines to make his work easier. His technical interest led to a mechanic's apprenticeship in Detroit, then a job at the Westinghouse Engine Company and, finally, a place in Thomas Edison's Illuminating Company.

However, it was Henry's obsession away from work at nights and on weekends that would lead to his fame. He designed his own internal combustion engine in 1891 and by 1896 had built his own car in the back shed. His boss and idol, Thomas Edison, asked to see the young engineer's handiwork, and told him not to bother trying to make it run on the electricity Edison was famous for.

At the age of 36, Henry left Edison to work at the Detroit Automobile Company, which went bankrupt before it could make any cars. But he persisted and, after attracting the attention of investors by winning a Michigan car race, he founded the Ford Motor Company in 1903.

Car design to a T

Henry was the majority shareholder in Ford, but his introduction of the inexpensive Model N in 1906 upset the other shareholders owing to the car's small profits. He was forced to develop an even cheaper model in secret. The result was the Model T Ford, also called the Flivver or Tin Lizzie, a sturdy, simple, practical and good looking car that was less expensive than other cars. At the relatively cheap price of US$950 the Ford company sold a record 10 000 Model Ts in its first year.

Henry was always looking for better ways to do things. For much of its production, the Model T only came in black because that colour paint dried faster than other colours. In 1910 he built a huge factory, where the Model T's metal body was punched out on the fourth floor,

Early days:
Wheely good ideas

Henry Ford was quoted as saying that history is bunk, but we'll give it a go regardless.

Around 5500 years ago, the Sumerians came up with an idea that changed the world forever – the wheel. But wheels always needed to be pulled by something with at least a couple of legs and a pulse. The paving of roads by the Greeks and Romans made wheels easier to pull, but pulling was still required.

In the 1400s, Leonardo da Vinci came up with an idea for a self-propelled vehicle, but he had so much painting to do it was never developed (just like many of his other crazy ideas, such as helicopters and bicycles).

In 1769 in France, Nicolas Cugnot designed the first automobile, a steam-powered tractor used to pull military equipment at the whiplash-inducing 4 kilometres per hour. In 1770 he built a three-wheeled carriage that could carry three passengers behind its steam boiler. Even at a snail's pace, in 1771 Nicolas managed to drive his steam-powered car into a stone wall, thereby grabbing the honour of causing the world's first car crash. By the early 1900s, more than a hundred manufacturers were building steam-mobiles.

Robert Anderson in Scotland invented the electric car in the 1830s. Electric vehicles were built in the 1880s, and between 1896 and 1915 more than fifty manufacturers built 35 000 electric cars in the USA alone.

Internal combustion engines were developed in the 1800s that ran on various types of fuels. This set the scene for the petrol-driven engines and cars of Otto, Daimler, Benz and Ford. Steam, electric and petrol engines competed for a number of years before petrol became dominant in the 1910s.

painting was done on the third floor and assembly was concluded on the second floor before the completed car rolled past the first floor offices. In a country where the average annual wage was around US$600, the price of a Model T dropped to $575 by 1912 and eventually to $280.

Moving right along

In 1913, Henry introduced the moving conveyor belt assembly line. He had been inspired by watching butchers stand below moving meat hooks, carving off cuts of meat until there was nothing left. Henry's workers reversed the process, standing their ground while cars in various stages of assembly moved past them. This process led to the Model T being assembled in just ninety-three minutes, much faster than the twelve hours it had previously taken.

The proportion of American cars that were Fords rose from about 10 per cent in 1908 to nearly 50 per cent in 1914. In 1914, 13 000 Ford employees made 260 720 cars. All other American car companies combined made 286 770 cars, but they needed 66 350 workers to do it. Three-quarters of the 15 000 cars imported to Australia in 1917 were Model Ts. Between 1908 and 1927, when Model T production ended, 15.5 million had been made.

Ford's you beaut ute

The car introduced new lifestyles and rules to the world. London installed gas-lit traffic lights in 1868; Cleveland, Ohio, fitted the first electric traffic lights in 1914. In 1905, the Automobile Association was founded in England to warn of police booking motorists travelling faster than the 32 kilometres per hour speed limit of the day. The first motorway, the M1, opened in England in 1959.

New industries appeared in countries around the world. In 1933, an Australian farmer wrote to Ford's Australian factory in Geelong to ask for a vehicle that was suitable for 'taking the family to church on Sundays' and for taking 'my pig to town on Mondays'.

In response, Lewis Bandt, a 23-year-old engineer who had joined Ford in 1929, designed the first coupe utility, or 'ute', which rolled off the production line in 1934.

Lewis took his design to the American Ford factory, where Henry Ford asked what the vehicle was. Lewis apparently replied that it was called a kangaroo chaser. Utes are now made and sold around the world.

In a tragic end to the story, Lewis was killed in 1987 while driving a replica 1933 ute. He collided with another car in rural Victoria while on the way to filming a documentary about his invention.

You're kidding:
It'll never catch on

When cars were introduced to Australia, you would have been forgiven for thinking it was just an overnight fad. When everyone else was selling their horsedrawn carriages, one man was buying them up at bargain prices and storing them in his shed in Geelong. Although never reaping the financial rewards imagined, it was a reasonable business decision considering some of the impediments in the early days of the auto industry.

In 1830s England, carriages that could carry twenty passengers were fuelled by coal, burnt to convert water into steam. The noisy vehicles emitted billows of smoke and ash, and spat out scalding steam and the odd bit of burning coal. In South Australia, steam-powered cars were so unusual that you needed to obtain permission from the mayor of Adelaide to operate one in the city streets. It appeared that steam-powered cars wouldn't become a global phenomenon.

In 1865, England introduced the Red Flag Act to warn people of an approaching steam car. A person was required by law to walk in front of a car waving a red flag during the day or a red lantern at night, while blowing a horn. Pedestrians would have plenty of time to get out of the way, because this ridiculous vision would pass at the legal speed limit of just over 6 kilometres per hour in the country, or 3 kilometres per hour in built-up areas. In 1896 the red flag rule was dropped and the speed limit was lifted to 20 kilometres per hour, and then 32 kilometres per hour in 1903, but horses still seemed like a more sensible option.

In 1879, American lawyer George Selden applied for the US patent for a petrol-powered car. For years, automobile manufacturers had to pay him 1.25 per cent of their annual sales. Fed up, Henry Ford took George's organisation to court, which, in 1911, pronounced that George could only obtain royalties on cars that resembled the long out-of-date 1879 vehicles. This finally freed up the American car market.

HOW IT WORKS:
Friendly fire

Steam cars burnt fuel such as kerosene to heat water in a boiler mounted in front of the driver – outside the engine, which is why it was called an external combustion engine. In automobiles such as the American Stanley Steamer, the pressure created by the expanding steam would rotate an axle. Although its simplicity was an advantage, there was a major disadvantage – if you think your car takes a while to warm up, imagine how long it took for a steam car's water to boil. Electric cars avoided this by using a battery to drive a small motor that turned a driveshaft.

Most cars today use the easily-burnt, high-energy liquid fuel, petrol. An alternator feeds electricity to spark plugs one at a time. A spark ignites the fuel, which has been mixed with air drawn into the carburettor and compressed, to cause a small explosion in a hollow cylinder. The number of cylinders depends on engine type. In each cylinder, the explosion pushes a piston, which turns a long crankshaft around. The crankshaft is connected to an axle via a driveshaft, which rotates the wheels.

Dr Rudolf Diesel patented a new engine in 1892, lending his name to the new fuel used to drive the engine. In 1971, Ralph Sarich in Perth invented a smaller, more efficient and less polluting orbital engine, drawing on ideas for rotary engines that had existed since the 1800s. As greenhouse gases increase and petrol availability decreases, electric engines are having a resurgence in the form of hybrids. In future, car engines are likely to be fuelled by hydrogen fuel cells.

SUCCESS SCALE

IMPACT

Massive. In 1900 there were 11 000 cars in the world. This wasn't many, considering that, thirty-five years later there were still 20 000 horses being used for transport in London alone. Today, more than half a billion cars are in use around the world – about one-third of these in the USA, the most car-dependant country. Another 30 million new cars are built around the world each year.

MARKETING

Car races and racetrack demonstrations were a common way to demonstrate early cars. Henry Ford organised black-tie car clinics in New York, and rodeos in the west where cowboys in Fords roped cattle.

MANAGEMENT SKILLS

Henry relied on instinct more than meticulous planning. When other car companies were moving towards more expensive cars made by skilled craftsmen, he wanted to build a car for the masses. He was described as prickly, eccentric – and brilliant.

FINANCES

Henry Ford became America's richest man. Like a number of other inventors, the success went to his head and he ran for the Senate. He lost.

References

Internet

DaimlerChrysler, at <www.daimlerchrysler.com>.
How stuff works, at <www.home.howstuffworkscom>.
<www.inventors.about.com>.
Technology in Australia 1788–1988, at
<www.austehc.unimelb.edu.au>.
Wikipedia, at <www.wikipedia.org.
Encyclopaedia Britannica online, at <www.britannica.com>.

Texts

Gross, Daniel, 1996, *Forbes Greatest Business Stories of All Time*, John Wiley and Sons, New York.

Inventor of Fish-Propelled Boat Makes Millions

How home sales demonstrated the value of innovative containers

Almost certainly you own some. You have certainly heard of it. But what is likely to surprise you is that this multimillion-dollar-making product is named after its inventor.

This enterprising person was born in 1907 in New Hampshire. Let's dub him Earl – because that was what he was called. Middle name, Silas. That won't help you a bit. His sur-

name gives it all away. So we'll seal that fact for the moment.

Earl grew up on various farms in central Massachusetts. His mother, Lula, took in laundry and ran a boarding house to supplement the money that his father Ernest made from the farm.

The 21-one-year-old Earl was an ambitious young man – no surprise, given that he appears here. 'I am going to make my first million before I am 30,' he announced.

Smoke and mirrors

To this end, Earl invented, designed and concocted furiously, filling his notebook with numerous products that he hoped would drag in the dollars. There was, for example, his distinctive comb design, dubbed the 'boudoir hair rake', the sport school comb complete with mirror and ruler and a dagger-shaped comb. The date written on the comb page of his notebook is 13 February 1937. The 13 February is unimportant. It's the '1937' that tells us that Earl was now 30. The absence of a boudoir hair rake in twentieth century business lore tells us that Earl would need to come up with something grander to make his – now post-thirtieth birthday – million.

Earl regularly wrote detailed descriptions of his inventions, posting letters to companies throughout the USA in the hope that one would say yes and shower him with riches. But it wasn't to be. To make matters worse, a tree surgery business that he established in the late 1920s was forced into bankruptcy in 1936 by the Depression.

Fortunately, Earl soon found a job in a nearby town at the DuPont plastics manufacturing plant. This was a fine environment for him. Later he would recall that it was at DuPont 'that my education really began'. After just a year, Earl bought some moulding machines and began his own company, making plastic beads and plastic containers for cigarettes and soap.

The Second World War years were profitable for Earl's business. He managed to acquire several defence contracts to make parts for gas masks and navy signal lamps.

In 1945, with the war over, Earl focused his attention on the burgeoning consumer demand for household products. Plastic today comes in all shapes, colours, sizes and finishes. But that wasn't the case in 1945. Plastic products were notorious for being brittle, smelly and greasy. Earl's first real claim to fame is that he developed a way of purifying black polyethylene slag, a waste material from petroleum refining, into a material that did not leak, was flexible, tough, non-greasy and semi-transparent.

Earl's company made cigarette cases, an unbreakable bathroom tumbler and various kitchen products, including a 'wonderli-

Early days:
Earl's family tree

Earl loved to tinker, and what better place to tinker than a farm. Young Earl designed and crafted all sorts of tools to help run the business. He even gained a patent for a frame that helped prepare chickens for sale.

Legend has it that the entrepreneurial lad discovered that lugging poultry and produce from door to door was a great way of increasing sales, far better than sales at the local market.

Earl worked on the farm throughout his school years, scraping through his high school final year in 1925. He then tried his hand at a number of jobs, including postal clerk and railway labourer. He also studied courses by correspondence.

In 1928, he completed a course in tree surgery and set up his own business, with a sideline in landscaping. The history books refer to this as Earl 'branching out'.

er bowl'. Plastic food containers were lighter and less likely to break than their glass and crockery counterparts. They were also cheaper.

Lifting the lid

Earl's second claim to fame is that he adapted the airtight, watertight lid used on paint cans to the plastics industry. It was the Tupper seal – Earl Silas Tupper's place in product history was assured. The Tupperware Plastics Company advertised that it had a product that enabled food to be stored for longer periods without drying out, wilting or losing its flavour.

Polyethylene, the versatile, cheap, flexible, durable and chemically resistant material that gives rise to innumerable everyday products and to Tupperware parties worldwide, is a simple substance. It contains only long chains – polymers – of carbon atoms, every one flanked by two hydrogen atoms. The chains are tens of thousands of atoms long – each snuggled up to another chain, making an impenetrable mass. Nothing but carbon and hydrogen. Simple in construction, but not so simple in creation.

Polyethylene comes from petroleum, or crude oil. This complex mixture of carbon-rich liquids, gases and solids that fuel our twenty-first century lifestyles, has a fair amount of history to it. Tiny, ancient, single-celled plants, such as blue-green algae, that floated around in prehistoric seas, lakes and rivers, died and accumulated within underwater sediments. Some biology happens, some chemistry, some physics. A couple of hundred million years later the petroleum is ready to be mined, processed and transformed into a Thatsa™ Bowl ('Unique thumb-loop handles ensure a superior grip and reduce fatigue when mixing').

An employee from the early days recalled in a television documentary that Tupperware was a unique product. There simply wasn't anything else like it. People were shocked that you could put food in this new container and it would keep longer and better than anything else. As a way of preserving food, Tupperware was better than wax paper or a damp cloth. Some claimed it was even better than a refrigerator.

The company shipped its products, now in a range of bright colours, to hardware and department stores across the nation. There was a *Home Beautiful* magazine feature: 'Fine Art for

39 cents'. However, sales lagged. Shoppers couldn't comprehend how to operate the containers. The novel lids needed to be depressed – 'burped' – to eliminate air before being sealed.

This was a product that needed to be demonstrated. Once people saw Tupperware in action, they were hooked. The stores weren't selling much, but, oddly enough, two sales representatives in Detroit were.

Decades later, one of these two reps, Gary McDonald, would describe the phone calls he received from Earl's first sales manager, incredulous as to what Gary was doing to sell such great quantities of Tupperware.

The consummate salesperson

The other Detroit rep was an ambitious, friendly young woman, Brownie Wise. She had been selling cleaning aids and brushes in homes in Detroit and had introduced Tupperware to homemakers – predominantly women – who gathered at a hostess's home.

A typical Tupperware party began with, 'Haven't you wished for unspillable containers that wouldn't break? I'm here to show you modern dishes for modern living that will save you time and money.'

Brownie was astute. In the 1950s, most women didn't work outside the home or own their own car. Going to a Tupperware party was fun. It was a way of meeting new people and perhaps getting away from the children for an hour or so.

Earl Tupper was enormously impressed by the sales that Brownie was making in people's homes. No doubt recalling his childhood door-to-door selling success, Earl Tupper invited her to visit him in Massachusetts. Soon Tupperware had a new vice-president and head of sales.

It was a good combination – the engineer and inventor, Earl Tupper, and the astute salesperson, Brownie Wise. In 1951, the Tupperware company removed all merchandise from store shelves, relying from then on entirely on direct home sales.

Tupperware parties proved remarkably successful. In 1951, there were eight distributors. Five years later, more than one hundred distributors throughout the USA sold products such as the Party Bowl, the Pie Taker and the Dip'N'Serve Serving Tray. Some distributors made millions.

Brownie Wise helped transform Tupperware Home Parties Inc. into a multi-million-dollar company. A great motivator, she was fond of saying, 'If we build the people, they'll build the business.'

There is an old black and white publicity photo of Earl Tupper and Brownie Wise that provides an unintended insight into the workings of the company. Earl and Brownie are pictured standing behind a bench. The balding Earl wears a check jacket, dark tie and white shirt. Brownie wears a dark, buttoned-up jacket over a white top. On the bench are small, white plastic beads – the raw material of Tupperware products. Earl and Brownie both

You're kidding: Fishing for fame

Wherever he went, Earl carried little pads of paper so that he could jot down ideas, inventions and uplifting sayings. His notebooks – preserved to this day – contain crude drawings and descriptions of the items that just might transport him to millionaire status. There is the no-drip cone to assist in the orderly eating of icecream, pants that won't lose their crease, the sweetie picture buckle, which allows you to adorn your belt with a most-loved photo, a novel way of removing a burst appendix and – our favourite – the fish-propelled boat. Take one fish, strap it beneath a boat . . .

HOW IT WORKS:
To burp or not to burp

While Tupperware products are well-designed and attractive, the key to their marketing, especially in the early years, was the burp. The deft thumbwork on the Tupperware lid, ejecting air from the container prior to sealing, was the key to differentiating Earl's products from those of his competitors.

Does the Tupperware burp make a difference to how long food lasts? Paul set out to investigate this in his extensive laboratories – the shelf in his kitchen – with a slice of Wonderwhite white bread cut into four 2 centimetre by 2 centimetre squares.

One square went into a small Tupperware container, duly burped. Another found its home in an identical container, sealed but unburped. The third bread square he placed into a sealed glass jar. The last square sat on the shelf exposed to air, moisture and air-borne microbes.

At this point we must admit to being somewhat sceptical about the value of the burp. Keeping bugs, oxygen and water away from food will help preserve it. However, you would need to be optimistic – or on the Tupperware payroll – to believe that expelling a small volume of air, say, 1–2 per cent, from the container would keep the Wonderwhite fresh and appealing, with few slabs of green mould.

Salespeople who worked for Tupperware during the 1950s later admitted that much mileage was made from the burp 'technology'. Ever-wilder claims were being made about the capacity of Tupperware to keep food for weeks, even months. It wasn't so, they knew; they just got carried away. And so did crates of the belching receptacles.

Back to the bread. A day later – no change. Two days, three days, four days, still no change. Those little squares of bread were just sitting there. Bugs might be breeding, but there was no outward sign. That Wonderwhite bread must incorporate Wondermould suppressor. A week passes. Finally, on day eight, a small green patch has appeared on the bread square in the glass jar. The bread exposed to the world has a fine, white coating on it, but no colourful bits. Now the interesting part. We open the Tupperware containers, burped and unburped. A pungent smell comes from each. The bread looks the same as the exposed version – fine, white coating but no colour.

Our conclusion – keeping bread out of the air is a good idea. That exposed bread square probably dried out quickly, whereas the one in glass had enough moisture to sustain green mould. No difference though between the squares in burped and unburped Tupperware containers. Tests more exhaustive than this one would be needed for us to be more definitive about the preserving properties of the burp. One thing's for sure, though – eating any of the squares now would certainly generate more than a burp.

hold a pile of the beads. You can hear the photographer saying, 'Hold up those little balls, Mr Tupper, Miss Wise … Look happy, please.'

But there's a striking difference between the poses that the two adopt. Earl's right hand is stretched out. The action looks forced. He has a nervous, half-smile on his face. You'd guess he is thinking, Let's get this damned photo shoot over with. But he also appears to be looking at those beads with the pride of an engineer who has come up with a smart process for making a new, versatile plastic container. There were challenges and difficulties along the journey, but he has solved them.

Contrast that with Brownie. She enthusiastically clutches a huge, overflowing pile of beads with both hands. She is enjoying having her photo taken. Her smile is wide-mouthed. She looks excited.

In those white beads, Earl seems to see plastic feedstock for his production line. Brownie, on the other hand, sees Tupperware parties, housewives delighting in their purchases, distributors to be encouraged and new ones to be brought on board. Earl has the backroom skills, Brownie the people skills. It proved to be a compelling combination for years and helped propel Tupperware to great success.

Eventually, Earl and Brownie would fall out. Perhaps this was inevitable, given their personality differences – the dour, conservative businessman and the bright and vivacious motivator and salesperson. But thanks to the two, Tupperware is a household name. Today, every 2.5 seconds somewhere in the world, a Tupperware party starts. More than $1 billion worth of Tupperware products are sold each year, across more than 100 countries.

SUCCESS SCALE

INNOVATION

Bright, attractive applications of plastic, with a distinctive sealing mechanism.

MANAGEMENT

Excellent businessperson + excellent salesperson + innovative product = good business.

MARKETING

What better way of allowing people to visualise Tupperware in their home than an inhome Tupperware demonstration?

References

Internet

About.com website, at <www.inventors.about.com>.
Filmmakers Collaborative/Blueberry Hill Productions 2003, *American Experience Presents Tupperware!*, WGBH Educational Foundation, transcript at <www.pbs.org/wgbh/amex/tupperware/filmmore/pt.html>
Tupperware home page, at <www.tupperware.com>.

Revolution is a Revolutionary Idea

The path to success was clear for James Dyson's invention

If necessity is the mother of invention, frustration must be its father. How many times have you been let down by a product, your exasperation at its failure inspiring the thought that you could probably make a better version of it yourself? It is just this type of frustration that led to the invention of one of the twentieth century's most revolutionary products.

In 1974, while renovating a farmhouse with his wife Deirdre in the English countryside, a young inventor by the name of James became increasingly frustrated with a wheelbarrow. The wheel became bogged in the mud, the ground was being carved up like a BMX track, steering a heavy load was difficult and the barrow would sometimes tip over.

James realised the solution was to replace the wheel with a large plastic ball, and hence the Ballbarrow was born. The Ballbarrow's trough was rustproof, tough and non-stick. The ball enabled it to travel in any direction, over broken ground and without leaving marks on the lawn. It was stable and could be pushed over mud without sinking.

However, James's invention was rejected by every garden and building shop he approached, so he resorted to placing a hand-drawn advertisement in the Sunday newspapers. This really got the ball rolling, so to speak, and the Ballbarrow became the market leader in England within three years.

But bear with us, for this story is not about the invention of the Ballbarrow, revolutionary as it may have been.

Cleaning up

By 1978 production was going well, except in the room where the finish was sprayed on to the Ballbarrows. It was here that the air filter would always get clogged with powder particles. The problem was similar to one that James had noticed at the farmhouse while vacuuming, when the filter bag on the vacuum cleaner would clog with dust and James would become frustrated by the loss of suction.

Maybe the same solution in his factory could be used to make a better vacuum cleaner, thought James … James Dyson. 'Most people don't think about the design of a frustrating object or how they can change it,' James told us. 'They just kick it in the hope that it will work on the second attempt. I think it's important to constantly challenge existing designs – there is always a way of doing something better.'

His idea for a new kind of vacuum cleaner well and truly eclipsed his Ballbarrow invention, as well as other, less successful, inventions of his, such as the portable, water-filled lawn roller called the Waterolla. Remarkably, Dyson vacuum cleaners and products have now achieved sales of more than A$7 billion around the world, and James and his company are worth millions.

But how did a clogged, factory air filter lead to a vacuum cleaner that introduced the biggest breakthrough in domestic cleaning technology since the invention of the vacuum cleaner in 1901?

James was born in 1947 to two British school teachers in Norfolk. While studying ancient Greek history, painting and classics at London's Royal College of Art he stumbled across the excitement of engineering. After working for seven years at an

Early days:
Hoovering history

To understand how big a revolution James Dyson's vacuum cleaner was when it appeared in the 1990s, it helps to understand how little the technology had changed in nearly a hundred years.

The first suction vacuum cleaner was invented by Ives McGaffey in his basement in Chicago. Called the Whirlwind Sweeping Machine, it was manufactured and first sold in 1869. It was made of wood and canvas, and its operator had to operate the suction fan by cranking a hand pump.

In 1901, British engineer Hubert Cecil Booth invented the powered vacuum cleaner. The petrol-driven unit was the size of an entire room, and had to be parked outside a building on a horsedrawn cart while cleaners vacuumed inside using hoses fed through the windows. The new technology was such a novelty that in the first years of the twentieth century wealthy people threw vacuum cleaner parties, at which guests sipped tea and lifted their feet while Booth's cleaners went to work.

James Spangler invented a more useful portable vacuum cleaner in 1907. In 1915 the first cylinder vacuum cleaner was invented by the company that later became Electrolux. Through the twentieth century only minor modifications were made to upright and cylinder vacuum cleaners, including the introduction of reusable paper bags and an automatic cable rewind.

James Dyson's purple and pink G-Force cleaner went on sale in Japan in 1986, and, with its price tag of roughly $3000, became a status symbol. The income enabled James to further develop his ideas. By 1993 he had perfected the Dual Cyclone bagless vacuum cleaner. In 1995 he introduced a cylinder model to complement the upright model.

James Murray Spangler, a cleaner in Ohio, invented the portable electric vacuum cleaner in 1907. He attached an old fan motor to a soap box, stapled it to a broom handle and used a pillow case as a dust collector. He improved the upright design and obtained a patent for his contraption in 1908.

Spangler's patent was then bought by his cousin's husband, whose name was William Hoover. So if it wasn't for James Murray Spangler's family connections, people today may well have been doing the spangling instead of the hoovering.

The word 'Hoover' has become more than a brand name; it has become part of English as a noun and a verb. But if it is up to James Dyson, in future we won't be hoovering or even spangling – we'll be dysoning.

engineering company that made high-speed military landing craft, he quit to devote time to his own business.

This brings us back to the Ballbarrow and the jamming spray painter. James designed and built a cyclone tower, which he had heard successfully cleared the air at a nearby sawmill. The tower alleviated the need for an air filter in the Ballbarrow finishing room by spinning powder particles to the outside of the cyclone where they could be removed and re-used. It didn't take James long to see that the same solution could be applied to make his home vacuum cleaner do without its filter bags, which only performed best with a new, unclogged bag.

Persistence pays

In a repeat of his experience of shops turning his Ballbarrow away, large appliance manufacturers rejected his ideas. 'Instead of the enthusiasm I had naively expected at such a revolution in vacuum cleaner technology, I received a sceptical and short-minded reception from the companies I approached,' James told us. 'One after the other, the large multinationals turned my idea down.' One company executive in Peterborough claimed in 1982 that, 'This project is dead from the neck up.' So, once again, James had to launch the machine on his own.

Brendan Coutts, a management consultant from the Hawthorne Academy, says this is typical of many organisations. 'There are always lots of great ideas floating around within organisations,' he says, 'but for various reasons they don't come up. This could be because employees are afraid they'll be seen as stupid, or that they can't question the existing way of doing things.'

Brendan says the same principle can apply for people providing ideas from outside the organisation. 'Management often doesn't like having their current way of doing things questioned. This can limit opportunities.'

Fifteen years and 5127 prototypes later, James had produced a bagless vacuum cleaner that he felt was ready for market. In 1993 the Dyson Dual Cyclone upright vacuum cleaner went on sale in England. Two years later it had become the number one selling vacuum cleaner in the UK. Today, Dyson is the best selling vacuum cleaner in Western Europe. The Dyson company sells around 3 million

vacuum cleaners each year in thirty-seven countries.

The company opened a research and development centre in Malmsbury, Wiltshire, in 1995, where James works with more than 1000 staff, including 350 other scientists and engineers. In 2005 Dyson invested more than $100 million in Research, Design and Development. As well as further developing the bagless vacuum cleaner, the Dyson company has produced a new kind of washing machine that has two drums to wash faster and better. Amazingly, the company is now working on a prototype robot cleaner that guides itself around a room as it vacuums. 'I can't say what's coming next though,' he told us. 'Sorry! Watch this space...'

So what advice does James have for entrepreneurs wanting to start their own business? 'It's

HOW IT WORKS:
Whipping up a storm

There are two types of traditional vacuum cleaners: upright and cylinder. Both have a fan that sucks air into a hose. The air draws dirt – and sometimes small, irreplaceable items inadvertently dropped beneath furniture – through the hose into a bag, where they are trapped. The bag allows air to pass through tiny pores to maintain airflow. Without the air holes, there would be no suction, the bag would blow up like a balloon and eventually pop. But dust can clog the pores, blocking airflow and reducing suction.

Dyson vacuum cleaners operate without bags. Instead, these new cleaners feature two cyclone chambers that cannot clog with dust so the vacuum cleaner never loses suction.

Air is drawn into a pipe and around a cone shape, which accelerates the air faster and faster. By the time it reaches the dust inlet, the air is spinning in a 1500 kilometres per hour whirl of air, or cyclone. That's about 35 000 revolutions per minute – much faster than a washing machine's drum at 1000 rpm, and even faster than the engine in Michael Schumacher's Ferrari, which ambles along at 19 000 rpm.

Particles, even microscopic ones, press against the outside of the cone when exposed to these huge forces. One cyclone chamber spins out larger particles of dirt and debris. Once these large particles have been removed, an inner chamber accelerates the air to remove smaller particles, even as fine as smoke particles. All the dust, dirt, hair, fluff and smaller particles are collected in a transparent bin, which James incorporated despite industry and marketing advice. Surprisingly, people want to see what is being sucked out of their carpet, perhaps so they can check for those small valuables sucked from under the furniture.

important to have ideas,' James says. 'But this is not enough. You'll need sheer dogged determination and perseverance to push the idea through. This is not a career path for the faint hearted'.

James says it is equally important to protect new ideas and inventions with patents, a lesson he learnt early on following a patent infringement battle with Hoover, which he eventually won and received $10 million in damages. So don't be a sucker. Follow James' advice and you might just clean up, too.

SUCCESS SCALE

MANAGEMENT
James treats his workers like family and is no stranger to the factory floor. With about 1000 employees, we guess he doesn't like working in a vacuum.

MARKETING
The Dyson does not rely on extensive advertising, instead basing its success on the quality of the product and word of mouth.

INNOVATION
An original idea from a man full of original ideas. With the Ballbarrow, Waterolla, vacuum cleaner and washing machine behind him, it seems he won't stop coming up with new inventions.

References

Internet
Dyson vacuum cleaner, at <www.dyson.co.uk>.
James Dyson, at <www.inventors.about.com>.
Science Museum, at <www.sciencemuseum.org.uk>.

Cleaning Up from Cleaning Up

Barb de Corti has a hand in cleaning up the environment

The average person spends six months of their life on the toilet. This fact makes an extremely odd start to this story, we admit. But it does make you think. Which is probably something you do a fair bit of during those six months.

Now think about this. Amazingly, it turns out that your toilet seat is likely to be remarkably clean. Far cleaner than anywhere near a kitchen sponge or dishcloth, which are likely to be a veritable zoo of germs. This is because microbes adore moisture. Toilet seats are simply too dry to support large microbe colonies.

But if you elect not to serve hors d'oeuvres à la dunny, you'll need to do a bit of cleaning. To keep cunning coliforms and other nasties at bay and to get the premises looking half-decent, at least when visitors are expected, the average person spends around thirty minutes every day vacuuming, dusting and cleaning the bathroom. Half an hour each day, over the average adult lifespan – kids do no cleaning – equals more than a year of scrubbing, wiping and whining. With going to the loo, that's a year and a half accounted for.

To help us battle the bugs, manufacturers offer a bewildering array of cleaners, scrubbers, solvents, deodorants and polishers. There are liquids, pastes, powders and spray packs. If you look under the kitchen sink or in the laundry, chances are that you'll find your own mini laboratory. The cleaners do the job well, but taming your environment may just be harming the wider environment. Pollutants are released when the chemical cleaners are produced. You might be allergic to some, or develop allergies through regular exposure. Bathroom mould removers can irritate and damage your throat and eyes.

Planet Ark ('Your daily guide to helping the planet') has this to say about common phenol-based disinfectants: Extremely toxic when swallowed and are rapidly absorbed by the skin; repeated exposure can damage the brain, liver, kidneys and spleen; kills the bacteria in sewerage-treatment works.

Working holiday

This is where Austrian-born Barb de Corti enters the picture. 'In 1994, the family took a holiday to visit my mother-in-law in Austria and she was raving about cloths that could clean using only water. No detergents or chemicals were needed,' Barb told us. 'We returned to Australia with some samples that she had given me.' Barb had discovered cleaning cloths produced by ENJO, the Austrian company established by Johannes Engl. She decided that the products might work well in Australia.

Barb invested $40 000 from her savings to buy the Australian franchise. She had a personal reason for her interest in chemical-free cleaning. As a youngster, her son Mark had asthma, an ailment Barb thought may have been exacerbated by the chemicals from cleaning products.

'I persevered with it until I found something that worked. I thought, if it helps my child I want others to have the same benefits.' She began by selling the cloths to the homeowners of Perth.

ENJO products do reduce chemical use and waste. They are

Early days:
Gloves off

In 1985, Friedrich Engl, an entrepreneurial man with years of experience in the textile industry, saw the effects of an oil spill on tranquil Lake Constance, a large freshwater lake bordered by Austria, Germany and Switzerland. Friedrich decided that there had to be a more environmentally sensitive mop-up process than the one relying on chemicals. Five years later, he had developed a remediation technique using textile fibrous tissue to soak up spilled oil.

Friedrich and son Johannes soon saw other applications of fibre cloth cleaners. This was the birth of microfibre ENJO cleaning gloves.

On his first day spruiking, Johannes managed to sell seven gloves. The proud Johannes put his mother and girlfriend to work, sewing. Between them, the women ultimately made thousands of gloves. In 1990, Johannes Engl established the ENJO company in Austria to market his products. The name ENJO is simply a contraction of Engl and Johannes.

Those first seven sales must have made a huge impact on Johannes. Like a number of entrepreneurs before him, he decided that a direct distribution system made the most sense. ENJO products represented such a departure from the existing way of cleaning, that it had to be seen – and felt – to be believed.

In 1992, with the introduction of new cleaning products, ENJO swept through Denmark, Norway and Iceland. The next year, the company cleaned up in Sweden and Central Europe. Today, ENJO's slogan is 'Clean the world' and, in most parts of the developed world, you will find ENJO.

expensive, but consumers could see that before too long there would be savings in saying farewell to Mr Muscle, AJAX and Pine-O-Cleen.

Entrepreneurs need drive, energy and enthusiasm. Barb de Corti has these in spades. She has turned her original $40 000 investment into more than $60 million. That's a lot of cleaning cloths. She is one of Australia's richest women.

Looking ahead

We asked Barb whether her Austrian background had helped in her dealings with ENJO International? 'No, not really. German language can be of assistance, but English is spoken in Austria.'

We also asked two obligatory questions. What do you most enjoy about your work? 'Working with people and making a difference,' Barb replied.

What advice would you give to people who are interested in starting a new business? 'Make sure you are absolutely certain you believe in the product and what you are doing and the rest will follow.'

Thanks to Barb, Australia is by far the most successful market for ENJO. Some 1500 consultants sell the environmentally friendly cleaning products at ENJO parties across the country. Barb has her eye on doubling this number and significantly increasing the ENJO business in Australia. With her grit and determination (to remove grit), Barb and ENJO are set to clean up their competitors.

What does the Australian Consumer Association, publisher of *Choice* magazine, have to say about ENJO? 'If you're keen to cut the use of chemicals in your home the ENJO Bathroom Glove, Kitchen Glove and Marble Paste are probably worth investing in. You'll need to establish new cleaning patterns, but in the long term it should save you time, water and money on cleaning products.'

The magazine does go on to somewhat spoil the ENJO sales party by saying that cheaper microfibre cloths tested work just as well as some of the ENJO products.

For the past year or so, Paul has been gloving up and cleaning the Austrian way. Although sceptical at first, he admits that the product works well.

HOW IT WORKS:
Secrets of the fibre

Dr Peter Dingle is a researcher at Perth's Murdoch University. He specialises in air quality in homes and office buildings, toxic chemical use, and air pollution and health.

'Our research shows that the average Australian home has twenty to forty chemical containers somewhere inside. Nearly all of these are highly toxic and most unnecessary,' Dr Dingle says.

ENJO claims to remove the need for up to 90 per cent of potentially harmful household chemical cleaners.

'Fibre technology for cleaning, such as ENJO, is much more efficient and safer. It doesn't pollute the air with toxic chemicals,' he told the ENJOhouse journal.

ENJO also says that their cleaners reduce use of water by 50 per cent, cleaning rubbish by 90 per cent, energy and money by up to 50 per cent and your expenditure of time by 50 per cent. We like the sound of the last one.

How does ENJO work? We contemplated undertaking exhaustive examination of books, scientific literature and the internet to bring you the secret of the ENJO fibre. Much easier though is to simply quote from the ENJO website. There, entitled 'The Secret of the ENJO Fibre', is a document that reveals all.

Every surface is uneven and has pores. Dust particles gather on and cling to the surface because of their fat share. The particular mechanics of the ENJO fibres uses the combination of microscopically small fibre strands with edges, barbs, knobs and capillaries. The ENJO fibre cuts the layer of fat, extracts the dirt from the pores and holds the dirt in the network of the fibre.

As for what the gloves are made from, ENJO 'processes fibres of polyester, acrylic, polyamide and cotton for its cloths and gloves with partly multifunctional qualities'. We think they're saying that ENJO cloths are made mainly from synthetic fibres.

'ENJO fibres are high-tech products,' says the website. To keep the tech under control, the cloths come colour coded. There is the white fibre, 'especially developed to remove lime residues and other obstinate soils from hard and smooth surfaces'. Perhaps this should have been lime coloured. Then there are cloths in green, orange, yellow, dark blue, salmon pink, yellow/white (this is getting complicated), black, pink and blue. If you are considering a career as an ENJO demonstrator, you would have to know your shades of pink.

You're kidding:
Dr Germ

Professor Charles Gerba of the University of Arizona's Department of Soil, Water and Environmental Science has a career that has gone to the toilet. Literally. He is an expert on domestic and public hygiene. Even though his department covers so many specialised disciplines, you'd think it could squeeze in just one extra – the Department of Soil, Water, Environmental Science and Toilets. Now that would be a memorable name. Unsurprisingly, Professor Gerba's nickname is Dr Germ.

You'd be hard pressed to come up with any gags that are new to Dr Germ. 'I've published several toilet papers,' says the bearded, bespectacled professor.

So he has. 'Chemical and microbial characterization of household graywater' is one of our favourites. Just four sentences in and you find yourself immersed in coliforms, faecal coliforms, faecal streptococci, *Staphylococcus aureus*, *Pseudomonas aeruginosa* and coliphages.

Professor Gerba and his colleagues set out to test microbe levels in a variety of domestic locations. They examined choppingboards, sinks, tap handles and floors, as well as performing extensive investigations, of a scientific nature, in toilets.

Probably the germiest item in your kitchen – the germiest room in most homes – is the sponge or dishcloth. This is where, as the researchers expressively say, 'faecal bacteria from raw meat festers in the damp, nurturing environment'.

As the good professor says, 'If an alien came from space and studied the bacterial counts, he probably would conclude he should wash his hands in your toilet and crap in your sink.'

```
┌─────────────────────────────────────────────────┐
```

SUCCESS SCALE

INNOVATION
Back to basics when it comes to cleaning, with environmental positives.

MANAGEMENT
Energetic and inspirational.

MARKETING
Consultants provide just the hands on, gloves-on demonstrations needed. Envirofriendly is a bonus.

THREATS
Cheaper microfibre cloths.

References

Internet

ENJO online (updated July 2003), at <www.enjo.net> and <www.enjo.com.au>.

Dr Germ, University of Arizona Alumni Association, at <www.uagrad.org/Alumnus/w05/germ.html>.

PlanetArk, Fact sheet on cleaning products and methods, at <www.planetark.com/products/PACleaningProdFactSheet.pdf>.

Choice magazine online, at <www.choice.com.au>.

On Your Bike

Bike designer shifts up a gear for racing success

Wilbur and Orville Wright achieved immortality for an event that lasted just 12 seconds above the dunes of Kill Devil Hills in North Carolina. On 17 December 1903, Orville was the first person to pilot and control a powered aeroplane.

Ten years earlier, the mechanically inclined brothers had been running a shop, selling and repairing bicycles. They had invented a self-oiling bicycle wheel hub and had built bikes of their own. It was the income from the bicycle business that funded the experiments that led to that famous first flight.

Orville's average speed during his 12 seconds aloft was just 11 kilometres per hour. You could have easily beaten him on a Wright brothers' bike. In fact, since bicycles have been built, people have been trying to beat others on them.

By the 1890s, there were organised bike races in many countries. In 1896, Frenchman Paul Masson won the Athens

Olympics, a third of a kilometre track race, averaging 50 kilometres per hour. Somewhat further, the first Tour de France was held in 1903, with sixty riders grinding their way over 2500 kilometres in nineteen days. There were six stages, with riders pedalling day and night to average 25 kilometres per hour for the journey.

Throughout the twentieth century cyclists continued to ride faster, thanks primarily to developments in engineering and in fitness and conditioning. (We'll overlook here the clandestine use of developments in the pharmaceutical industry.)

In 1928, Willy Hansen of Denmark won the Amsterdam Olympic 1 kilometre track trial on a bike that today you wouldn't think twice about pitching at the local tip. Willy's average speed was 48 kilometres per hour. In Athens in 2004, Australia's Shane Kelly blitzed the field in the kilometre track trial at an average of 59 kilometres per hour.

Superbike

One of Australia's most significant contributions to bicycle development began in 1992, almost by accident. Lachlan Thompson, an aerospace engineer and lecturer at

You're kidding: Back peddling

You peddle them, but they're very different from your average bike or even a Superbike. Speedbiking involves propelling yourself along flat roads in aerodynamic capsules. Designs differ – in some you even travel backwards – but the principle is the same: minimise air resistance.

If the idea of travelling at high speed while lying in a carbon fibre capsule centimetres above the ground and pedalling furiously appeals to you, then head for the flat Nevada desert. There, Sam Whittingham, a 32-year-old Canadian, reached the world record for self-propelled speed: 130 kilometres per hour.

If that doesn't impress you, in October 1995, professional cyclist Fred Rompelberg hit 269 kilometres per hour at Bonneville Salt Flats, Utah, by riding in the slipstream of a lead vehicle.

RMIT University in Melbourne, was organising props for a photo shoot. In perhaps a subconscious nod to the Wright brothers' bicycle–aviation connection, the idea was to contrast the sleek, modern lines of a jet fighter with the clunky engineering of a street bike.

Needing a bicycle, Lachlan wandered out into Swanston Street, where he chanced upon Kathy Watt, a young cyclist. Kathy volunteered to contribute to the photo and asked if she could have the aerodynamics of her bicycle and riding position tested in RMIT's wind tunnel. This was a truly fortuitous meeting. Later that year, Kathy would win gold and silver medals at the Barcelona Olympics.

Lachlan, also fortuitously, had an interest in bicycle design. Here was the perfect opportunity to develop his ideas. With Olympic cycling team coach Charlie Walsh, and the Australian Institute of Sport,

Lachlan established a team to develop a superbike. The superbike had to represent the finest in engineering and materials, but had to be suitable for mass production.

The rules at that time stipulated that racing bikes needed to have a triangular frame configuration, so that's what the team began working on. However, soon after they started, the Union Cycliste Internationale relaxed its requirements for three-tube frames, thereby enabling the engineers, scientists and students in the Superbike team to demonstrate their flair and ingenuity.

Nineteen ninety-three was spent creating mathematical models of performance for various designs and materials testing. There was much work undertaken with elite cyclists in the wind tunnel and out on the track.

'You become very tied to the athletes' performances,' says Lachlan. 'It's an almost symbiotic

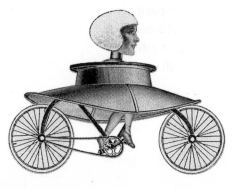

HOW IT WORKS:
First Kathy, then 200 more watts

Given its lightness and strength, carbon fibre technology was an obvious choice for the Superbike frame. The wind tunnel tests pointed to a streamlined, aerodynamically shaped, single hollow shell for the Superbike, instead of the traditional tubular frame. Carbon fibre handle grips attached directly to the wheel forks replaced handlebars. The Superbike weighs just 5 kilograms.

Dr Ian Crouch of the Cooperative Research Centre for Advanced Composite Structures (CRC) at RMIT says the secret is applying state-of-the-art aerospace technology to the design and manufacture of competition bikes.

In a CRC media release, Ian explains: 'We looked at each element of the frame and modelled the stresses and loads applied to it. We then chose the best combination of carbon fibre grades and fibre orientation, for each individual element – the same basic approach as in the aviation industry.'

With tubular-framed bikes, some of the rider's exertions go into twisting and flexing the frame. The Superbike's shell is particularly stiff, allowing almost all of the rider's energy to be directed to the pedals. The stiffness, coupled with the bike's streamlined shape and lightness, reduces the power needed to match the speed of a conventional tubular-framed racing bike by at least 5 per cent.

Testing the frame on athletes showed that they were each averaging an extra 200 watts of energy output on the road.

relationship and you get a fantastic buzz. You work away and think to yourself, This could win, but until you're actually there by the track and you do it there's always that doubt.'

Once the Superbike team had perfected its design, someone was needed to produce the bikes. The team chose Salvatore Sanonetti. Salvatore knew bikes. He was an Olympic cyclist who had competed for Australia in the 1976 Montreal Olympics. He also knew material technology, as he ran a company that made injection-moulding dies for major car manufacturers.

Salvatore headed Bike Technologies, which manufactures Superbike frames. Camp-agnolo, an Italian company, produced the rest – wheels, cranks, pedals, seat and chain.

The award-winning Superbike carried cyclists to medals at Olympic and Commonwealth Games, as well as in various international track championships. Bad news came in 1997 when the Union Cycliste Internationale banned single-shell designs, such as the Superbike, from elite competition. However, Superbikes are still being made for triathletes and sports and recreational cyclists.

Early days:
Gearing up

The earliest machines that could be said to be the forerunner of the modern bicycle appeared in the 1800s. Baron Karl von Drais de Sauerbrun of Germany invented a two-wheeled wooden running machine. Riders perched on a seat between the wheels, clutching a steering bar attached to the front wheel while pounding their feet along the ground for propulsion. Fred Flintstone did much the same thing in his Stone Age cartoon car.

A Scottish blacksmith, Kirkpatrick Macmillan, is credited with inventing the first bicycle with foot pedals. He achieved this in the early 1840s, cycling 64 kilometres to

---

Early days cont.

Glasgow in 1842 to demonstrate his new device. No known drawings of the bicycle exist, but Kirkpatrick probably pushed treadle-type pedals backwards and forwards, supplying power to the back wheel via rods.

Other outdoor types in Scotland and elsewhere came up with modifications on the design over the next two decades. As better designs allowed the *vélocipède de pedale* to go faster, a swifter name was needed for the new machines. In 1868, the word bicycle became popular throughout Europe.

Soon pedals were being fitted to the front wheel. Rubber tyres and wire-spoked wheels helped begin to remove the boneshaker label often given to older wood-spoked wheels and iron rims. The larger the front wheel, the faster your pedalling would take you. Hence, the penny-sized front wheel – up to 150 centimetres in diameter – and tiny farthing-sized rear wheel of the penny-farthing bicycle. These bicycles could go fast, and so could their riders, head over heels from their precarious positions high above the front wheel. So the 1870s and 1880s saw the introduction of safety bicycles, with a low centre of gravity, similar sized wheels, the rear driven by a chain and direct steering through the front wheel. A major breakthrough in design during this era of bicycle boom was the replacement of hard rubber tyres with air-filled pneumatic tyres, which helped to cushion bumps.

In the early 1900s, bike makers introduced three-speed hub gears. By the 1920s, gears that lifted the chain from one sprocket to another – derailleur gears – were de rigueur.

Despite what you may have heard, Leonardo da Vinci did not invent the bicycle. Story has it that manuscripts examined by Italian monks in the 1960s included a sketch of a pedal-propelled bike. Scholars now believe that a modern-day forger added a few lines to a couple of circles drawn by Leonardo, turning them into what appears to be a bicycle.

References

Internet

Encyclopaedia Britannica online, at <www.search.eb.com/eb/article-230020>

Olympic cycling medalists, at <www.hickoksports.com/history/olcycl.shtml>.

Powerhouse Museum, Sydney, at <www.phm.gov.au/hsc/bike/>.

RMIT Openline, at <www.fsae.rmit.edu.au/Media/Openline%20August.pdf>.

Superbike, at <www.phm.gov.au/hsc/bike/features.htm>.

What's This 'ere?

Hear! Hear! The bionic ear

A 17-month-old baby boy, Callum, plays with a toy as his mother discusses his situation with a doctor. From behind Callum's ear, a wire protrudes from a microphone connected to a transmitting device on his head. The transmitter is held in place by a magnet, which sits with a receiver inside Callum's skull, having been surgically implanted there two weeks earlier.

Callum seems unaware of the discussion. In fact, he doesn't hear a word of it because he was born unable to hear almost any sound from the outside world, a condition known as profound deafness.

Every year, three in every 1000 children in Australia are born deaf or develop loss of hearing before they learn to speak. About 1 million Australians have some degree of hearing loss, and 20 000 of them are profoundly deaf. Worldwide, about 120 million people have significant hearing loss, so they and those close to them may be interested in Callum's appointment with the doctor.

The doctor switches on the implant, which beeps and buzzes

to life. Callum looks up, startled, his eyes enquiring what the doctor, his mother, and the ABC radio journalist witnessing this amazing event are doing. Next, the doctor switches on the microphone, and Callum, for the first time in his life, hears his mother's voice.

'Hello darling, can you hear mummy?' Callum's eyes widen like saucers. His mother says that she loves him. And then he starts to cry.

Callum was just one of the tens of thousands of recipients of a cochlear implant, also called a bionic ear. Until 30 years ago, a bionic ear was the stuff of fiction. Many doctors thought profoundly deaf people would never be able to hear or learn to understand speech. But today, 'Can you hear me?' are perhaps the first words heard by thousands of people around the world who have gained or regained their hearing, thanks to the implanted device. And the device exists thanks to the tireless commitment of Australian surgeon Professor Graeme Clark.

A quiet start

Graeme decided when he was 10 years old that he wanted to be an ear, eye and throat surgeon, a more original choice of career than his friends who, like most primary school students, aspired to be train drivers, fire fighters or movie stars. But Graeme wasn't just trying to be original – his motivation came from the fact that his father had been profoundly deaf his entire life.

In the late 1950s, after studying medicine and surgery at the University of Sydney, Graeme headed to Edinburgh to specialise in ear, nose and throat surgery. He returned to Melbourne to practise surgery but realised there was little that he or anyone could do for profoundly deaf people, whose hearing organs were too damaged to be helped by hearing aids.

Then, in 1966, Graeme read a scientific paper describing an American surgeon's success in

Early days:
Sounds like a good idea

Italian scientist Count Alessandro Volta (of voltage fame) was the first person to artificially stimulate sounds using electrical impulses in the human ear – his own. Although he wasn't at all deaf, it is surprising he didn't become so when in 1790 he inserted two metal rods into his ears and connected them to a battery. He described the sound as a boom inside his head, followed by the sound of bubbling soup. Don't try this at home.

People have known that sound was a vibration since Greek philosopher Pythagoras (of triangle fame) worked it out in the sixth century BC; and transmission of sound by nerves to the brain was understood 1800 years ago. The parts of the inner ear, including the cochlear, were discovered in the 1500s, but it wasn't until the 1920s that Hungarian physicist Georg von Bèkèsy explained the inner workings of the cochlear, which won him the Nobel Prize for Medicine and paved the way for the development of the cochlear implant.

For the past several decades, hearing aids have partially overcome some forms of hearing loss. But it wasn't until the late 1950s that the first profoundly deaf people began to have their acoustic nerves directly stimulated by electrodes. In one early case in 1957, a patient described hearing background noise and sounds like a roulette wheel and a cricket. Not great, but it was a start.

Early implanted hearing devices had only one electrode, which rarely led to understandable speech. The bionic ear developed by Graeme and his team, while not producing sounds perfectly, can lead to adults understanding 80 per cent of speech. Graeme and his colleagues continue to improve the device for people from older than 80 to younger than a year, in the important language-learning stage of life.

using electrical stimulation to enable a profoundly deaf person to experience hearing sensations. Inspired, the following year Graeme began research on his own; he continued with limited funds and a slowly growing team at the University of Melbourne. He overcame criticism from scientists who said a bionic ear couldn't work, and that it would be dangerous for the patient. He also had to overcome the challenge of fitting a bunch of wires into the tiny, delicate, spiral space of the inner ear, to which is connected the smallest bone in the human body, the 3 millilitre stapes, controlled by the body's shortest muscle, the 120 millilitre stapedius.

Breaking the sound barrier

As is so often the custom for genius, the solution came to Graeme while on holiday. He was sitting on a beach pushing blades of grass and twigs into spiral shells. He noticed that he could reach furthest into the shells, which had a similar structure to the inner ear, using grass that was stiff at one end while thin and flexible at the other. This was how he could solve the problem of reaching the electrodes far enough into the small space of the inner ear to pick up speech frequencies.

In 1978 his first prototype was installed inside the ear of a man named Rod Saunders, who had lost his hearing after a head injury. Graeme and his colleagues were devastated to find that the device failed at first, but they soon realised this was due to a loose connection in the external equipment rather than any insurmountable problem with the technology.

Over the following six years, twenty-two adults received implants. In 1985 the bionic ear passed the all-important US Food and Drug Administration tests and became the first multi-channel implant to be approved. The first child, a 5 year old, received an implant in 1985.

More than 55 000 people in 120 countries have now had the device implanted. The ability to connect a technological device to the central nervous system is considered to be the most significant advance in helping profoundly deaf children to communicate since the introduction of signing at the Paris Deaf School 200 years ago.

HOW IT WORKS:
Safe and sound

The visible outer flap of the ear – much more visible on some people than on others – is one of three parts of the ear. The satellite dish-like outer ear collects sound and funnels it through the ear canal. As part of the middle ear, the eardrum vibrates when hit by soundwaves, and the vibrations are carried and amplified by tiny lever-like bones and muscles to the inner ear's cochlear, from where the vibration of fluid sends messages to the brain.

The bionic ear replaces the function of the entire ear, capturing sounds and electrically stimulating the nerves inside the inner ear to create hearing sensations. For a person wearing a cochlear implant, a small microphone held behind the ear sends information to a speech processor kept on a belt or in a pocket. The speech processor picks up the sound and converts speech into an electrical signal. This is sent via a transmitter held behind the ear by a magnet to a receiver that has been surgically implanted inside the mastoid bone in the skull (the implant operation takes between ninety minutes and five hours under general anaesthetic). There are no wires connecting the inside and outside components. The receiver decodes the signal and sends a pattern of electrical currents through an array of twenty-two electrodes in the inner ear's cochlear. The currents stimulate the auditory nerve fibres that send messages to the brain.

Talk isn't cheap

In the face of early dismissal of his ideas, Graeme found it difficult in the 1970s to raise the funds required to develop the bionic ear. His big break came when Melbourne television station owner Sir Reginald Ansett organised a telethon. In a subsequent telethon, activists cut the phone lines – there is some opposition to implants in the deaf community, some of whom do not see deafness as a disorder that needs a cure.

A thousand person years of work and $30 million worth of research later, bionic ears are a real commercial success. The Cochlear Corporation, which was developed as a result of Graeme Clark's work, now serves the vast majority of the tens of thousands of people around the world using a bionic ear. The market is nowhere near saturated, with only about 10 per cent of people who could benefit from the bionic ear having had one implanted. The device costs around $20 000, with another $12 000 of medical and rehabilitation costs. Cochlear's shares rose from $2.50 in 1995 to more than $48 in 2001 and in 2006 were above $50. However, Graeme's Bionic Ear Institute receives no royalties from cochlear sales so he is once again short of funds.

Graeme has received many awards, including the $300 000 Prime Minister's Science Prize and Australia's highest civilian honour, the Companion of the Order of Australia. He has been named Senior Australian of the Year, and has been elected to the Royal Society and the Australian Academy of Science. Some awards come with a healthy monetary prize, and the Australian government has provided some funds. But without a source for the $40 million he estimates is required for further research, he may need to look for another telethon.

You're kidding:
Bionic body parts

Some people do not have functioning nerve cells, which means that they cannot use cochlear implants. So the researchers who developed the bionic ear have invented a new hearing aid called the Tickle Talker. This delivers signals to a deaf person's hand, which tickles them using electrodes worn in a glove.

A bionic eye may soon join the list of bionic facial features. Researchers from the University of New South Wales and Newcastle University have developed a tiny camera attached to a pair of glasses, which radios an image to a microchip fitted in place of the eye's lens. The microchip sends the image through wires to the retina, which passes the image to the brain.

The researchers tested the bionic eye by implanting it into a sheep, which was then able to tell the difference between light and dark – we assume it could therefore identify the black sheep of the flock. The next step is to improve the detail the sheep can see, and then trial the technology in humans.

SUCCESS SCALE

Value of product
Life-changing, but there is some criticism from within the deaf community, some of whom do not see deafness as a disability.

Producer
The Cochlear Corporation controls most of the market. Two other companies, Advanced Bionics in the USA and Med El in Europe, also manufacture and distribute multichannel cochlear implants.

Management skills
Graeme's excellent scientific skills combined with his sympathy as a medical specialist for human feelings has made him a well-liked leader of the 100 or so cochlear researchers who have worked with him.

Finance
A lack of funding is not due to lack of trying – for example, Graeme offered his Prime Minister's Science Prize funding to research if twenty-five other Australians would contribute a similar sum (sadly, they didn't).

References

Internet

Australian Academy of Science Nova, at <www.science.org.au/nova>.

Australian of the Year, at <www.australianoftheyear.gov.au>.

Bionic Ear Institute, at <www.bionicear.org>.

Cochlear Corporation, at <www.cochlear.com>.html>.

National Academy of Sciences' Beyond Discovery, at <www.beyonddiscovery.org>.

A Problem to Sink Your Teeth Into

Recaldent becomes a solution for recalcitrant dental patients

Paul was early for his appointment with the dentist, an appointment he was looking forward to. Eric Reynolds is no ordinary tooth puller. He is Dean of the Dental School at the University of Melbourne.

The waiting room was the old boardroom of the dental school. In the middle of the room was a long wooden table surrounded by eleven orange-cushioned chairs. On the wall hung huge portraits of former dental school deans, adorned in academic finery. Not a tooth was visible in any painting!

On Paul's arrival, Katherine Nesbitt, the super-efficient communication officer, presented him with samples of the product invented by Professor Reynolds. There was Recaldent grapefruit-flavoured lozenges, Recaldent chewing gum and melon-flavoured Recaldent tooth mousse ('topical creme with bio-available calcium and phosphate').

Eric strode into the room. He wore a navy blue suit, blue shirt and a red tie with a blue stripe. On his lapel was a small gold V signifying the prestigious $50 000 Victoria Prize, which he had just won. He had also recently received an Order of Australia for service to dental health.

A tall man with dark hair and a friendly face, Eric has a touch of charisma that makes you wonder what he is doing in a dental school.

Milking the research

Eric's research into products that decrease tooth decay began in the early 1980s.

'There was a connection between the School of Dentistry and the Dairy Board. At that stage, milk was being criticised because of the fat level and it was cut from schools.'

The Dairy Board was concerned about the criticism and asked the dental school to look into the evidence that milk is good for teeth. Researchers have known since the 1940s that dairy products reduce tooth decay (caries, in dentist speak).

'In the mid to late 1980s, we had a large team working on this,' explained Eric. 'All the research pointed to the fact that

the protective nature of dairy products was related to the protein and the calcium phosphate. The research indicated that the proteins were very important, so we patented that.'

Say cheese

Dairy products ward off caries predominantly through a chemical process involving calcium. The calcium in milk is in a specific form that is stabilised by casein, the major protein in milk. Add acid to milk and casein precipitates, forming curds, from which cheese is made. Add whey, and hey, you have Miss Muffet's favourite meal.

Drink milk, eat cheese and your stomach's own digestive enzymes will rip the casein apart, creating short protein fragments called peptides. One group of peptides seeks out calcium and phosphate, transforming them into a substance that our bodies, including our teeth, can use.

'Research in the 1990s was undertaken to understand the interaction between the peptide and the calcium and phosphate,' said Eric.

'We found that under certain conditions you get clusters of calcium phosphate that come together and form little nan-

Early days:
Fluoride fights fillings

Arguably the greatest advance in dental health has been the addition to water supplies of small amounts of fluoride. Does fluoride protect teeth? Paul was born before water was fluoridated. His parents began giving him fluoride tablets when he was three or four years old. His younger sister Caroline had the tablets from an early age. Today Paul has more than a dozen fillings; Caroline has hardly any.

If you are not impressed by a sample size of just two, in the 1930s and 1940s, American researchers stumbled upon the fact that people living in areas where there were naturally high levels of fluoride had fewer caries than people elsewhere. Just 1 gram of fluoride added to 1000 litres of water is sufficient to decrease the number of decayed, missing and filled teeth in children. Stop the fluoridation and soon far more children are being asked to open wide and told that this won't hurt a bit.

Fluoride concentrates in the dental enamel and in bones. It increases the resistance of tooth enamel to erosion by acids in the mouth. Toothpaste manufacturers have known for years the advantages of fluoride, which is why it is a common additive.

Around 70 per cent of Australians drink fluoridated water. Most Australians have had water fluoridation for 25–50 years. Today, every capital city except Brisbane has water fluoridation.

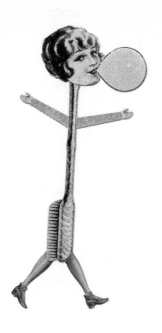

phate is leaping onto your teeth. This is a good thing.

'The nanoclusters will penetrate quite deeply into tooth enamel. You can actually produce crystal growth in the enamel that is very like what happens during the normal tooth formation process.' Eric illustrated this by holding up a before-and-after photograph of a hapless set of brown-stained teeth magically transformed by diligent Recaldent gum chewing into a Miss World winning set of gnashers.

'You are remineralising the tooth. Where this is a real breakthrough is these little nanoclusters can diffuse into the porous enamel because they are so small. Once they come into contact with the tooth enamel they dissociate and release calcium phosphate. It ends up in enamel.'

In their experiments, the dental delvers have created the beginnings of tooth decay in extracted teeth. By applying their calcium phosphate, they can completely repair the decay.

'Won't lowering tooth decay put your students out of business?' Paul asked.

'Fluoridation reduced tooth decay and was supported by the dental profession. The profession supports this product. The demand for dentists has

oclusters, about two nanometres in diameter.' A nanometre is one-billionth of a metre. In other words, pretty small.

Recaldent

Luckily, these nanoclusters are very stable. Ignoring the screech of drills from the dental students next door, Eric's team produced, filtered and dried the calcium phosphate particles in their laboratory. They christened the resulting white powder Recaldent.

Recaldent was soon being added to gums, mints and other products. Chew the gum and within seconds the calcium phos-

HOW IT WORKS:
A problem to chew over

Your mouth may contain more microbes than there are people on Earth. That's more than 6 billion bugs. Every time you eat, you are feeding your mouth bugs.

Your mutans streptococci, lactobacilli and other mouth microbes adore sugar and don't much care where they get it from – sweets, cake, soft drinks, even milk, fruit and vegetables. Sugar in, acids out, is the bugs' modus operandi.

The thin layer of bacteria that stick to the surface of your teeth is known as dental plaque. Brushing helps to remove plaque. However, it may build up, eventually forming a harder, more permanent layer called tartar, which is impervious to your toothbrush.

Tooth enamel is made up of calcium phosphate crystals. The bug acids, especially lactic acid, attack your teeth, depleting them of the essential calcium and phosphate. This demineralisation leads to tooth decay.

Protect your teeth by limiting the amount of sugar you eat or the number of times each day you eat sugar. Brush and floss, or at least rinse your mouth, after a sugary snack attack.

The good news is that there are ways of remineralising your teeth to replace the essential minerals lost. Saliva can help here, because it contains calcium and phosphate, so chewing sugarless gum is good for your teeth because it stimulates saliva.

Recaldent is particularly effective at remineralising teeth because it supplies copious quantities of calcium and phosphate in just the right form to be taken up by your teeth.

You're kidding:
My tooth feels fine now, thanks. Really.

The ancient Sumerians thought that tiny worms caused dental decay and toothache. Five thousand-year-old oriental lore says that toothache can be treated by inserting tablets containing garlic into the ear opposite the aching tooth.

The Romans were particularly fond of toothpicks made from vulture quills. About one thousand years ago, the Chinese invented the toothbrush. It had an ivory handle and bristles made from a horse's mane. Some eight hundred years later, the British constructed a toothbrush made of bone with bristles wired into hand-drilled holes.

You can drill back a long way to find out about the history of dentistry. Egyptian mummies have been found with teeth made of ivory, and with transplanted human teeth.

The earliest method for extracting a diseased tooth was holding a V-shaped piece of wood against the tooth and banging it with a mallet. Nice. Early Chinese dentists simply used their fingers, which they strengthened by pulling nails out of wood. The ancient Greeks and Romans used forceps and pliers and a host of other techniques you don't want to know about for tooth loosening and improved leverage. Frenchman J. C. de Garengeot modified the dental key, a device featuring a handle set at right angles to its shaft. You twisted and twisted – a process surely mimicked and amplified by the miserable patient – until the tooth emerged. Unfortunately, the root was often left behind.

People have known for a long time that decayed tissue must be removed from a tooth. Early devices for achieving this were small picks and scissors. The forefathers of your dentist – and, we are betting, his or her heroes – employed two-edged cutting instruments that they twirled between their fingers. In 1790, increased – there is a desperate shortage. There are more dental visits than ever before because there are many other things for dentists to do. There are cosmetic treatments and the need to tackle gum disease, which is a significant problem for adults.'

Recaldent, now patented worldwide, resulted from the partnership between the University of Melbourne, the Victorian Dairy Industry Authority and Bonlac Foods. Recaldent is now made in Victoria by Recaldent Pty Ltd using the milk protein, casein.

The product has been successfully launched in the large American and Japanese markets. Recaldent gum is Japan's second best-selling sugar-free gum.

Science is soon to make milk an even better product. Eric explains.

'We have added Recaldent to milk and showed that you can substantially increase the remineralisation potential.

Does Eric use Recaldent gum? Of course. 'I chew it regularly, my wife does and my daughter does.'

At 11 o'clock precisely, a woman appeared at the boardroom door. It was time for Eric's next appointment.

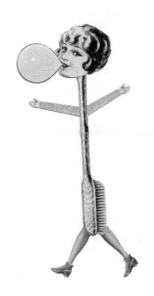

References

Internet

Dr Eric Boyden, at <www.drboyden.net/fun/history.html>.
Jane Austen Society of Australia, at <www.jasa.net.au/london/dentist.htm>.

Texts

Travers, B. and Muhr, J. 1994, *World of Invention*, Gale Research, Detroit.
University of Melbourne media releases.

Of Mice and Mould

How Howard Florey gave global health a shot in the arm

In the early days of the Second World War, a small team of scientists in Oxford, England, performed one of the most important medical experiments in history. They considered their work so urgent that they came in to begin the experiment on the weekend. On Saturday, 25 May 1941, the team injected eight mice with a lethal dose of streptococci bacteria. Four of the mice were treated with a new drug the team were developing, while four were left untreated, as controls.

By the next day, the treated mice had recovered while the others were dead. During a world war, the lives of eight mice may seem insignificant. But the rescue of half by the drug – penicillin – led to the treatment of Allied soldiers as early as D-Day in June 1944, and probably influenced the outcome of the war.

As the developer of penicillin, Adelaide-born Howard Florey is conservatively credited with saving 50 million lives since the Second World War. He had gathered a team of scientists at Oxford University in the 1930s to commence a careful investigation of the properties of antibacterial substances that are produced by mould. While flicking through a medical journal, team member Ernst Chain found an article written by Scottish scientist Alexander Fleming. This prompted the team to begin looking at penicillin.

A day after the historic experiment on mice, hundreds of thousands of troops were rescued from Dunkirk as the Nazis advanced towards the French coast. Fearing the Germans would overrun Britain, Howard's team prepared for the need to destroy the lab results. They hid the penicillin by injecting it into their clothes so it would be safe for future research.

Working holiday

Luck and war seem to surround the penicillin story. Alexander Fleming had been conducting an experiment with bacteria around the end of World War I when a tear fell from his eye into a culture plate laden with bacteria. He later noticed that a substance in his tear, which he named lysozyme, killed the bacteria but was harmless to the body's white blood cells. Years later in 1928, when Alexander returned from a two-week holiday, he found that a bit of common mould had fallen into a discarded culture plate containing bacteria, forming a clear patch. He recognised this pattern from his previous experience with lysozyme. He concluded that the *Penicillium* mould was producing a substance that killed the bacteria, so he named the substance penicillin.

Alexander's discovery was an amazing piece of luck. If he hadn't left a Petri dish of bacteria on his bench when he went on holidays, if he had properly disinfected the dish, if the weather had been different from the ideal conditions for bacteria and mould growth in the laboratory and, especially, if Alexander hadn't the experience to recognise the importance of the observation, penicillin may not have been discovered as an antibiotic.

But Alexander couldn't extract or concentrate the bacteria-killing substance, so he couldn't test it as a treatment for general infections. He published his results and then moved on to other research – leaving Howard and his team to discover penicillin's use as a lifesaver more than a decade later.

Early days:
Pricks and pain prior to penicillin

Three thousand years before penicillin, moulds and fermented materials had been used to cure various skin infections, although without an understanding of how they actually worked.

It wasn't until the late 1800s that scientific studies of antibiotics began. French chemist Louis Pasteur, after discovering that infectious diseases are spread by bacteria, observed that mould inhibited the growth of anthrax.

British surgeon Joseph Lister noted that samples of urine contaminated with mould didn't allow bacteria to grow, but he was unable to identify the substance in the mould.

French medical student Ernest Duchesne successfully tested a substance from mould that inhibited bacterial growth in animals, but he died at an early age in 1912, never seeing the world's acceptance and use of his important discovery.

Prior to the development of penicillin, a simple finger prick from a rose thorn could lead to an arm amputation, common eye infections could lead to blindness and razor cuts during shaving could be fatal.

Man or mouse?

The results of Howard's mice experiment in 1940 were so exciting; he knew that it was now time to test the drug on humans. In 1941, a 48-year-old London policeman, Albert Alexander, had a swollen face, eyes and scalp following a scratch by a rose thorn. He had already had an eye removed and abscesses drained, but even his remaining eye had to be lanced to relieve the pain of the swelling. As Howard's first patient, he was given penicillin and within a day began to recover. But Howard's team didn't have enough of the drug to see the patient through to a full recovery. Their efforts to recycle the penicillin by extracting it from Albert's urine failed: he had a relapse and died. Because of the awful experience, the team then concentrated their efforts on sick children, who did not

HOW IT WORKS:
No dividing = no multiplying

There are now more than sixty antibiotics, which are substances that fight bacteria, fungi and other microbes harmful to humans – the word means against (*anti*) life (*bio*).

Penicillin was the first naturally occurring antibiotic discovered. It is obtained in a number of forms from *Penicillium* moulds. Penicillin G is the most widely used form.

Bacteria reproduce by dividing to produce two new cells. They enlarge to about twice their size before the DNA chromosome is copied. The two new chromosomes move apart and a membrane forms between them.

Penicillin interferes with cell wall growth, so the new cell wall won't be able to form and prevents the cell dividing. The drug doesn't harm old bacterial cell walls; it just stops new ones forming. This means the bacteria can't reproduce, so the disease can't spread.

require such large quantities and showed more encouraging results.

In 1943, Howard travelled to North Africa to test the effects of penicillin on wounded soldiers. His trials were seen as a miracle. While it seems an obvious suggestion today, instead of amputating wounded limbs or simply leaving them to heal, Howard suggested that soldiers' wounds be cleaned and sewn up, and that the patients then be given penicillin. Thanks to Howard and his team, the drug was available to treat Allied troops by the end of the Second World War. It has since revolutionised medical science and saved millions of lives.

Saving millions but not making millions

At first, Howard's team made penicillin using old dairy equipment and hospital bedpans to grow mould. Liquid containing penicillin was drained from beneath the growing mould and filtered through parachute silk. But the team needed drug companies to help produce the large amounts required for patients.

Companies in England were unable to help out on a large scale because of the war, so Howard Florey took a small and valuable sample of penicillin aboard a blacked-out plane to the USA. The trip was against the wishes of Ernst Chain, who wanted to first patent their ideas in England. This would have made the team very rich, but instead Howard and Ernst, with Alexander Fleming, settled for the fame and small fortune of the Nobel Prize for Medicine in 1945.

Mould money

Many others became rich thanks to penicillin. At the end of the Second World War, the life-saving drug was worth a fortune on the black market. People would steal it from army hospitals, water it down and sell it in quantities too small to save a life. The penicillin black market is the background to the Graham Greene book and Orson Welles film, *The Third Man*.

Following his dangerous trans-Atlantic flight, Howard had explained his penicillin-making methods to the US government, which sponsored a crash program in making large quantities of the drug cheaply. Hundreds of scientists in dozens of laborato-

You're kidding:
Aspro – no pain, no gain

While Howard Florey allowed other companies to snap up penicillin patents, another Australian was more ruthless and managed to make a fortune internationally from an already well-established drug, aspirin.

Today, the equivalent of 100 billion 500 mg aspirin tablets are produced each year. But the use of aspirin is nothing new – around 2400 years ago, Hippocrates recorded the use of willow tree bark to make a pain-relieving powder. The active ingredient, salicin, was discovered in the 1840s, but it wasn't marketed as a drug until Bayer patented Aspirin in Germany in the late 1800s.

During the first World War, with the disruption of German supplies, the British government offered £20 000 to anyone who could come up with another way to make aspirin. George Nicholas, a pharmacist in Melbourne, developed a technique and named his product Aspro after Nichol **as Pro** ducts.

By 1919 his company was selling more than £4000 worth of pills a month. His brother Alfred, who had been working as an importer and grocer and so understood how to market the product in Australia and overseas, helped make Aspro an international hit; the Nicholas brothers became very wealthy. Not surprisingly, Aspro was never popular in Bayer's homeland, Germany.

ries were involved in the work. By late 1943, Pfizer scientists had developed techniques for mass production. The US government required most penicillin be dedicated to the war effort, and authorised nineteen other companies to produce the drug using Pfizer's technology. Increased production meant the drug became cheaper, with the price of a single dose dropping from almost priceless in 1940 to $20 in 1943 and 55 cents in 1946.

Following the war, several companies took out patents on penicillin production techniques, obtaining royalties from the increasingly widespread sale of the drug. Between the end of the war and 1950, the price of around half a kilogram of penicillin dropped further from about $4000 to less than $300. However, several strains of bacteria became resistant to penicillin after a few years. To overcome this problem, scientists in the 1950s produced artificial penicillin by chemically changing natural penicillin.

SUCCESS SCALE

VALUE
The invention of penicillin was miraculous; it saved millions of lives.

DISCOVERER
Alexander Fleming, who first identified its properties, is widely known for the discovery, but couldn't develop it into a drug.

PRODUCER
Howard Florey's team developed it into a drug at Oxford, but couldn't produce enough to meet even local demand. American companies produced enough of the drug to make it available to the masses.

FINANCE
The Oxford team did not take out a patent on the discovery and so missed out on a share in the multi-million-dollar industry they created. As the number of Nobel Laureates who can be attributed to one discovery is limited to three, many of Howard's team missed out, not only on fortune, but also on fame.

MANAGEMENT SKILLS
Howard Florey was an innovative manager, drawing together a team of scientists at a time when science was more commonly an individual pursuit. He also showed great leadership and even greater determination to see the production of penicillin grow to a large scale in the USA within four years of discovering its potential as a life-saving drug.

References

Internet

The Helix, at <www.csiro.au/helix>.
John Curtin School of Medical Research, at
<www.jcsmr.anu.edu.au/hon_roll/florey.htm>.
Pfizer, at <www.pfizer.com>.
Technology in Australia, at <www.austehc.unimelb.edu.au>.

Search for Success

How the Google search engine made nearly a googol dollars

What's the most popular colour in the world? Have a guess. Go on. Now try to guess what is the world's most popular animal, alcoholic beverage and flower.

Thanks to the internet and Google, we can objectively answer all these questions. Or at least tell you the most frequently occurring examples from each category on the web.

The most-mentioned colour on the web is red. When we looked, there were 580 million indexed web pages with one or more references to the word 'red'. Finishing second was blue, with 374 million hits. Green gives 357 million hits, brown 278 million, and yellow 199 million. Don't feel blue about us including the other use of the word blue in the total, it still rates strongly. Mix together red and blue – purple occurs 47 million times. Now let's get arty. Crimson has 15 million hits, magenta, mauve and cerise 7, 2 and 2 million respectively. We ignored black and white in this survey, as we recall being told at school that they aren't colours.

The internet's sought after animal is the cat, which claws around 10 per cent more hits than the dog. The most mentioned alcoholic drink is wine, and roses receive more hits than any other flower.

That you can keep tabs on the web with such ease and efficiency is due to two Stanford University computer science students, Larry Page and Sergey Brin. Larry and Sergey came up with the idea for Google in the mid 1990s. The name they chose for their internet search engine is a variation on the word googol, which denotes the numeral 1 followed by 100 zeros. Google, the company, has a mission of bringing order to seemingly a googol of web pages. Google opened for business in 1998.

Spiders order the web

To get a feel for the complexity of the web, imagine an almost endless library that has just been hit by Hurricane Dewey. Every book, brochure, pamphlet and flyer has whirled up into the air, around and around, to thud back to the shelves at random. The hurricane has also blown in supermarket flyers, party invitations, schoolchildren's homework assignments and photocopied material from the good, the bad and the ugly. Now, a horse manual in Italian sits next to *War and Peace*, a Mexican cooking book abuts *Amazing Science*, by, umm, Paul Holper and Simon Torok, and a Flat Earth Society tract lies on top of a manual for a tractor. A librarian's worst nightmare. How would you find anything?

The web is this crazy library. Internet search engines bring some order to the chaos. With a good search engine you will find a recipe for beef and bean burritos for your next dinner party or information on how to fix your tractor.

Google uses electronic 'spiders' to read and catalogue millions of web pages each week. Spiders are artificial intelligence programs that trawl the internet and report results back to huge databases. Every time you run a search, you are interrogating those databases.

The mission to control 1 followed by 100 zeros of internet information took a large step forward in 1998 with the establishment of Google Inc., supported by 1 followed by 6 zeros of US dollars. The location was Menlo Park, California.

Early days:
There's the rub

Larry Page and Sergey Brin first met in 1995. Larry, the son of a computer science professor, was born in 1972 in Michigan. A year younger, Sergey moved as a boy from his birthplace, the Soviet Union, to the USA with his family. Early in 1996, the two joined forces on a Stanford University research project to test their idea for a new approach to searching the internet.

In the early days of the internet, search engines had it easy. In 1994, there were just 100 000 or so web pages. It was relatively straightforward for software programs to track the available information by compiling a database of key words appearing on websites and then searching this information.

Larry and Sergey decided that a search engine that analysed the relationships between websites would produce better results than word count techniques.

Today, you could easily be saying, 'I'll BackRub that', rather than 'Google search', because BackRub was almost the name of Larry and Sergey's program. The name arose because their approach checked incoming links from other sites, known as backlinks, to estimate a site's importance. There is also an element of 'You rub my back and I'll rub yours', which linking to each other's websites does in a virtual way.

Determining the quality of a web page from the number of other pages that link to it is a key strategy employed by Google. The theory is that a website that many other sites are pointing to must be pretty good. There is some additional clever tweaking, with extra points for a link from a prestigious website such as Yahoo! home page.

Larry and Sergey presided over a company that was handling 10 000 search queries each day. Within months, staff numbers reached eight and daily searches 500 000. A few months later, the number of searches hit 3 million.

By mid 1999, Google had secured $25 million from Silicon Valley venture capital firms. 'We plan to aggressively grow the company and the technology so we can continue to provide the best search experience on the web,' announced Larry Page. By mid 2000, a partnership with Yahoo! led to 18 million queries daily. A free search facility brings in no direct revenue. Advertising does that. Hence, AdWords, an online advertisement program.

Searching millions, making millions

Our search for knowledge is insatiable. By 2001, Google was answering more than 100 million queries a day, from computer users and, for the first time, from mobile phones. At year's end, Google boasted that it had indexed 3 billion web documents.

In early 2004, Google reached a new peak, handling more than 80 per cent of all search requests on the web through its own and partner sites. That year, the company went public, offering shares at $85 each. This is what the public had been searching for – Google now had a market value of more than $23 billion. In an instant, many of Google's employees became paper millionaires. In nine years, Larry Page and Sergey Brin had turned an idea and a few thousand dollars into a massive enterprise that is almost indispensable to computer users the world over. By 2005, 1000 Google staff members were millionaires, at least in theory.

In the search for simpler searching, Google introduced a toolbar, which allows users to search directly from their screens without visiting the Google home page. Next came Google Zeitgeist, a strangely compelling tally of the world's most searched terms in a host of categories, updated weekly, with monthly and annual tallies.

Enter Google Image Search for tracking photos and drawings. To pay the bills for the rapidly-growing workforce and for necessities such as healthy meals prepared by the Grateful Dead's former chef, the company introduced Google Catalog Search. This program enables you to probe department store and other flyers on the web.

Other perks of working at the Googleplex company headquarters – apart from tracking, via Google, your escalating financial status – are hallways full of exercise balls and bicycles, workout rooms, a massage room, video games, foosball, a baby grand piano, a pool table and table tennis.

Google continues to introduce innovative new products, such as Google News, which accesses thousands of leading global

---

You're kidding: The internet – different of good grammar

Be wary about information on the internet. People would carefully check material before making it available to the whole planet, right? Wrong! There are 500 000 sites with 'congradulations'. No congradulations for their spelling. Any word that can be misspelt will be. Thirty-four thousand times, in the case of 'mispelt'.

The internet not only wreaks havoc on the language, it 'wrecks havoc' 25 000 times. Rather than 'to all intents and purposes', there are 670 'to all intensive purposes'.

English teachers insist on 'different from' (formal usage), rather than 'different to' (informal usage) and certainly rather than 'different than', which is American usage. The internet sort of agrees, with, respectively, 96 million, 7 million and 25 million instances. Intriguingly, there are 600 000 'different of's.

An easy way of comparing the popularity of different terms or even different people on the internet is offered by <googlefight.com>.

Egosurfing is hunting the web for your own name.

Googlewhack is a pastime in which you attempt to find two words that produce just one search result. This is far harder than you might expect. We gave up when 'Sudoku deoxyribonucleic' yielded more than 200 hits.

sources. The very aptly-named Froogle provides product searches. Gmail is a free email service, supported by advertising. Google Earth allows you to zoom in on satellite pictures of your neighbourhood.

With the web hosting more than 50 billion publicly available web pages, a tally that grows hourly, the need for Google is stronger than ever.

HOW IT WORKS:
Well red on the internet

There are rules of logic that govern how to search a massive database such as the internet. Nineteenth century Irish mathematician George Boole established a system of logic, known in his honour as Boolean logic, which dictates how search engines do their job.

Google returns only pages that include all of your search terms. When we searched for 'red' we got 580 million hits, 'blue' gave 374 million, but 'red blue', which Google takes to mean 'red and blue' yielded 118 million hits. In other words, Google knows of 118 million pages that include both colours. 'Red blue football' further narrowed the field to 12 million hits.

Word order matters in searches. 'Football red blue' elevates the position of the Adelaide Crows Australian Rules Football Club website compared with a search for 'red blue football'.

Google ignores common words such as 'a', 'the', 'where' and 'how'.

Perhaps the most useful tip of all is that you can search for a specific phrase by enclosing it in double quotation marks. 'Melbourne Football Club' renders 1.78 million hits; "Melbourne Football Club" gives 53 000 targeted hits. The latter search would, for example, ignore a web page that refers to the Richmond Football Club, which is based in Melbourne.

Adding 'site:au' will restrict the search to only Australian web pages. Five minutes on the internet will tell you that there is lots of rubbish and dubious information around. Using 'site:edu' will limit the hits to educational institutions, which may increase your chances of obtaining credible information.

By scanning Google's advice pages, you'll learn that there are many additional tricks for setting out searches. For example, type 'DVD player $350..400' and you'll get pages on DVD players that cost between $350 and $400. Particularly clever is the fact that if you place the tilde (~) immediately in front of your search term, you will also get synonyms. Thus, '~DVD player' gives you hits for many other items of electronic equipment, including MP3 players and CD players.

Google offers an Advanced Search link on the Google home page, <www.google.com>. By filling in boxes, you can be as specific as you like. You can, for example, search for websites written in the past six months that contain the 'Melbourne Football Club', mention either 'skill' or 'class' and don't mention the term 'mid-season slump'.

Finally, try typing into Google the following and observe the results: 'What is the population of Australia?', 'Who is the president of France?', '45x4', '5 Australian dollars in Chinese money' and '10 gallons in millilitres'. When we did this, Google offered the following answers: 20 090 437, Jacques Chirac, 180, 30.3335192 Chinese yuan and 37 854.118 millilitres, respectively.

References

Google home page, at <www.google.com>.
'Google Receives $25 Million in Equity Funding', Google media release, at <www.google.com/press/pressrel/pressrelease1.html>.
Wikipedia, at <www.en.wikipedia.org/wiki/Google>.

i Plays the Music

How digital music has wired millions for sound

One fine autumn day in 1995, some of the most popular bands from Australia and New Zealand were jammed into a small studio – to jam. The musos, including Neil Finn, gathered to play outside Australia's seat of government in Canberra. But the only way you could listen to Crowded House at Parliament House that day was from the comfort of your own house. The music and images emanating from this event did not flow directly to the ears and eyes of a live audience – they were sent to millions of people around the world by way of the web.

It was the first time in Australia, and only the third time anywhere in the world, that musicians had broadcast live over the internet. The virtual concert was a demonstration of how far state-of-the-art multimedia had come, and gave a glimpse of the future of music. But it was also a demonstration against being short-changed by the lack of copyright controls in the fast-developing digital music revolution – as music travelled the world free of racial, political and geographical boundaries, it also travelled free of charge.

Sure enough, just hours after the performance had started, an American college radio station used the music. In fact, anyone with an internet connection could obtain free music from the Australasian bands wherever they were – maybe silverchair from an office chair.

Line up online

Virtual concerts and online radio stations use a process called 'streaming'. You can listen to music as it arrives to a buffer on your computer without waiting for the entire file to be downloaded. Streaming is also a way of listening to a song before you decide to buy it online.

The purchase of music online is more than simply ordering CDs from online stores such as Amazon.com and awaiting their arrival in the post. It is about downloading songs over the internet. The digital music revolution is perhaps the biggest change to the way we listen to music since the introduction of the record brought music to the masses, or the subsequent displacement of records by compact disks.

The revolution takes the revolution out of playing music – no more spinning disks. Gone are the days of scratched LPs, worn out tapes and unattractive CD towers in living rooms. Digital music has enabled people to search for music they may not even find in a large music store. It has made it easy for people to distribute music easily and cheaply – even their own songs recorded at home. And artists can sell their music direct to fans without the need for a record label contract.

What a tangled legal web we weave

The concerns aired by musicians in Canberra in 1995 were justified, as pirated music and illegally copied audio and video files were shared around the world with the help of services such as Napster and the Australian-based Kazaa.

Many people may not even know they are breaking the law when they download copied music. Others know they're doing something wrong, but think of it as minor, as illegal as parking in a no-standing zone. Or some feel that free sharing of material was what the internet is all about.

Music theft is nothing new; bootleg tapes were available in the 1960s. The difference with downloading and burning to CD is the efficiency and multiplicity: an MPEG bootleg can be picked up and passed on by thousands of people online in a minute.

iPod therefore iAm

Rather than a company inventing and developing a product and then convincing the population of its need, the music technology industry saw that the audience was using digital music online and so produced a device that could make it portable. Enter the iPod.

Although the details are shrouded in secrecy, the iPod was conceived by Tony Fadell, who demonstrated his idea to Apple in early 2001. Initially, the company hired him as a consultant to build the iPod, and then employed him as the director of a 30-member team of designers, programmers and engineers, including iPod designer Jonathan

Ive. Apple chief Steve Jobs, the 50-year-old billionaire inventor of the Mac, also played a key role in the iPod's design and feel. The story goes that this may be why the iPod is louder than other digital audio players – apparently Steve is partly deaf.

The font used on iPod screens is Chicago, a throwback to the 1984 user interface of the original Macintosh computers, possibly the only other computer product that owners have loved like pets.

The original iPod was released in October 2001; subsequent models have helped Apple dominate the portable digital music player market. Apple has sold around 30 million players in its first four years, with sales increasing hundreds of per cent each year. An industry of accessories has grown up around the product,

Early days:
Around the world in 80 kilobytes

The internet was proposed in 1964 as a decentralised and therefore robust communication system that would survive a nuclear war. The first internet site was installed in 1969; by 1972 there were thirty-seven sites. The number of computers linked to the internet has been doubling each year since 1987 – hundreds of millions of people are now connected.

The world wide web, which is the graphical interface of the internet, was initially developed in 1989 by Tim Berners-Lee at the European Particle Physics Laboratory, known as CERN, as a means of sharing research and ideas effectively throughout the large organisation. The first piece of web software was introduced to the public at the end of 1990. A year later the amount of traffic flowing through the web was still only equal to about one floppy disk, but by the end of 1993 more than 50 000 files were being sent each month and traffic on the information superhighway was growing at more than 10 per cent per month. Fasten your seatbelts, because the amount of information transferred over the superhighway now doubles every year, and there are millions of websites on servers around the world.

Meanwhile, music had progressed a long way since Thomas Edison invented the phonograph in 1877. In 1948 Columbia Records started selling long-playing (LP) records that rotated at 33⅓ revolutions per minute, and a year later RCA introduced 45s. Next, in 1965, came eight-track magnetic tapes, and then cassettes, which, together with the Sony Walkman, made recorded music mobile.

These were all analogue devices, which means they used recordings of sound waves that were analogous – or similar in shape – to the original soundwaves. The groove on an LP, for example, is similar in shape to the pressure variation of the sound it recorded.

Digital audio uses a binary code (zeros and ones) to represent the sound. The first digital recordings, on compact disks, appeared in 1983; they were followed by digital audiotapes and minidisks.

Although the first MP3 players appeared on the market in 1998, the German Fraunhofer Institute had patented the format nearly a decade earlier in 1989, around the same time that the web was being developed in Switzerland.

including voice recorders, games and other software to enhance the iPod's usability, and hardware such as speakers, car kits and cases made of fur.

I tune into iTunes

iTunes, which opened in the USA in April 2003 and Australia in October 2005, is the world's largest and most popular of the 350-odd websites from which you can download music for your portable digital music player, computer, mobile phone or other device. By the end of the digital jukebox's first year, 20 million tracks had been downloaded by keen iPod owners willing to pay about a dollar per song. After two years, iTunes has sold 500 million songs.

In the $500 million a year pop singles market, internet downloads overtook CD and vinyl sales for the first time in 2005. Although distributors of classical music have been slower off the mark in selling online, downloaded music is not confined to pop songs. In fact, when the BBC website in the UK offered Beethoven symphonies free of charge in July 2005, 1.4 million tracks were downloaded, far outstripping even the highest selling pop hits.

But paying to download music is still in its infancy. The number of songs swapped illegally has fallen but it remains in the hun-

HOW IT WORKS:
Ripping, dithering and clipping

Just as a JPEG file greatly reduces the size of digital images without noticeably affecting quality, MPEG compresses music files. MPEG stands for Moving Picture Experts Group, which developed the compression system for video files. The part of the system used to compress sound is MPEG Audio Layer III, so music files names end with the suffix mp3. There are other music file formats, with acronyms such as WMA and WAV, but MPEG is the most common.

Copying music from a CD is known as ripping – not to be confused with ripping off, which is how the music industry sometimes sees it. A ripper copies a song from a CD to a computer hard drive. MP3 software then compresses it: a minute of music takes up about 10 megabytes of space in its decompressed form, but only 1 megabyte when compressed. This enables a large amount of music to fit into a small amount of space and reduces the time it takes to download a song from a couple of hours to just a few minutes.

Because they remove information that is not essential, compressed music files do not audibly reduce quality. This removal is comparable to removing punctuation and abbreviating words when sending a text message from a mobile phone – the meaning is not lost, although your grammatical credibility is. MP3 files remove sounds such as those inaudible to the human ear, or remove a quiet sound that has been drowned out by a simultaneous louder sound and is therefore redundant.

The quality is good enough for computer speakers or the small headphones used on portable music players. However, audiophiles complain that digital audio files don't sound great on a quality home stereo.

dreds of millions. And keep in mind that there are 10 thousand million songs sold on CD in the USA alone. Worldwide, there is about $45 billion worth of recorded music sold each year, but only 6 per cent of this is delivered digitally.

The International Federation of the Phonographic Industry predicts that digital music will account for one-quarter of the market by 2010. So will this, and performances such as the one in Canberra in 1995, mean the end of live music? We don't think so. After all, the e-book hasn't destroyed book sales. And the prediction of a television executive in 1950 that the televising of football matches would result in games being played to empty stadiums has proven not to be the case.

You're kidding:
iPod One

The songs saved on an iPod, and the list of favourites saved in what is known as a playlist, says a lot about people. The iPod display screen acts as a window into their musical tastes and – some say – their soul. Hence, it was of international media interest when, in early 2005, the playlist of one of the world's most famous people was revealed. Here's what US President George W. Bush listens to. His playlist is dominated by country music and singers of the 1960s and 1970s. Van Morrison and The Knack make an appearance, but there are no musical genres less than 25 years old. Analysts note the lack of variety: no gay or black artists and only one woman, Joni Mitchell. In total, George W. has about 250 songs – not fully utilising the 10 000-song capacity of the device given to him by his daughters.

Playlistphobia may become a new fear. If you want to avoid having your personality judged, keep your list of favourites a secret.

SUCCESS SCALE

VALUE OF PRODUCT

The digital music revolution has saved the pop single and, rather than ruining the recording industry, has revived it – combine digital and physical music sales and the industry is growing.

IMPACT

Huge, as seen on one side by the growing number of customers buying music online, and on the other by the number of record companies and bands selling their music online. In 2004, George Michael left the record industry to distribute his songs online for free – although we're not sure how many people were ever really that happy paying for it in the first place.

PRODUCER

Apple is the leader of the digital music revolution, with the iPod the most popular portable digital music player and iTunes the most popular downloadable music site.

MANAGEMENT SKILLS

A good move by Apple chief executive Steve Jobs putting iPod conceiver Tony Fadell on staff.

FINANCE

While initially looking like robbing artists of copyright royalties, online music looks like it will help the music industry. And iPod and iTunes haven't hurt the fortunes of Apple, either.

References

Internet

Answers.com <www.answers.com>.
BBC, at <www.bbc.co.uk>.
CNN, at <www.CNN.com>.
How stuff works, at <www.home.howstuffworks.com>.
iTunes, at <www.apple.com/itunes>.
MP3 Handbook, at <www.teamcombooks.com>.
Wired, at <www.wired.com>.

Text

The *Guardian* newspaper <www.guardian co.uk>.

Getting on Board

The future of tourism is looking up

On 30 April 2001 Dennis Tito took his seat at the beginning of his holiday flight. He probably acted as any one of the 3 million airline passengers around the world do each day before taking off. He buckled up in his seat, perhaps looked apprehensively at the exit, almost certainly felt excited about his trip. The difference between Dennis and other travellers was that, once the engines ignited, Dennis blasted off at more than twice the speed of sound. Vertically. Into space. Welcome to the holiday of Dennis Tito, the world's first space tourist.

For the 60-year-old American multimillionaire, 2001 really was a space odyssey. After taking off in April 2001, he spent a week orbiting Earth onboard the International Space Station. And today space tourism is literally taking off. On average over the past 35 years there have been only four space flights a year, mainly by the USA and Russia. In 2004 there were five, three of which were by a private company. Market research suggests that in the USA alone around 10 000 people would also pay up to $100 000 for a ride into space, so in future space flights may become a daily occurrence and a multi-billion-dollar industry.

On the verge

Mankind took a giant leap towards affordable space tourism in late 2004 when SpaceShipOne blasted more than 100 kilometres above the Earth's surface, ten times the cruising height of long haul flights. Five days later, on 4 October 2004, forty-seven years and a day after *Sputnik* became the first object launched into space, *SpaceShipOne* repeated its suborbital flight – into space but below the satellites that orbit our planet. The repeat feat earned the *SpaceShipOne* consortium the $10 million Ansari X Prize. The prize, for the first non-government spaceship to fly safely above 100 kilometres twice in two weeks, had been announced in 1996 by a non-profit foundation of donors to encourage private space travel. The prize echoed the $25 000 Orteig Prize for the first aeroplane that could fly non-stop across the Atlantic, which was won in 1927 by Charles Lindbergh and inspired massive growth in the aviation industry.

The key to affordable space tourism is a reusable space

Early days:
From pioneer to commonplace

Sea cruises around the world were once the domain of explorers who claimed new lands, but now people pay thousands of dollars to relax on an ocean-crossing ship.

Antarctica was off limits to all but the most dedicated scientists until just a few years ago; now, there are travel guidebooks for the icy continent.

Just as Henry Ford came up with a way to bring the invention of Karl Benz to everyone, space tourism is about making flights into space affordable.

vehicle to keep costs down. *SpaceShipOne* cost less than $50 million, which is one-fifth of the cost of a single Space Shuttle flight. The two *SpaceShipOne* flights were piloted by Mike Melvill and Brian Binnie, and funded by Microsoft co-founder Paul Allen. But it was the entrance of another billionaire, Richard Branson, that has brought flights in *SpaceShipOne* closer to all of us.

With similar aims to his Earth-based ventures, Richard Branson, through Virgin Galactic, plans to offer budget flights that are out of this world. With the designer of *SpaceShipOne*, Burt Rutan, he has formed The Spaceship Company and plans to build five new eight-seater spacecraft similar to *SpaceShipOne*, but larger and with more windows. As with Virgin Blue, free inflight meals are unlikely.

Virgin Galactic plans to have its first launch from a new spaceport in New Mexico in around 2008. You can also reserve a $250 000 seat with a minimum 10 per cent deposit today, but tickets are expected to drop to $100 000 once demand rises. NASA has predicted that, within a couple of decades, the cost of a seat on space flights may sell for as little as $20 000.

Honeymoon on the Moon?

While space tourism sounds exciting, a voyage to Venus or a journey to Jupiter is still science fiction. But Virgin Galactic flights will certainly be off the planet. Passengers will travel to the fringe of space, reaching the suborbital height of 160 kilometres above the Earth's surface. For a few minutes the weightless conditions will allow passengers to float over to a window for a cosmic view before they commence their return to Earth. At speeds of more than 3000 kilometres per hour, the flight will take less than three hours, but you are promised a DVD recording so you can relive the experience.

For those who can't wait until 2008 for a Virgin Galactic flight, Space Adventures now offers the first commercial flights to the limits of our atmosphere. For $10 000 you can take part in a series of zero gravity parabolic flights in an aeroplane. About three times this amount will get you onto a MIG-25 jet program that flies to the edge of space. A suborbital flight will set you back $138 000.

But why not go the whole hog today and buy a ticket on an 8-day orbital flight for $28 million?

In a deal brokered by Space Adventures, Gregory Olsen from the USA paid this amount for a ride in October 2005 on the Russian Soyuz spaceship to the International Space Station, where he spent a week 400 kilometres above the Earth's surface. He was the third space tourist, following American Dennis Tito's journey in 2001 and the journey in 2002 by the appropriately-named South African, Mark Shuttleworth.

Not for those with an atmosfear

There will be limits to who can be a space tourist. You won't be able to simply buy a ticket: health checks will be essential to ensure physical fitness (travellers with a fear of heights or a weak heart would be weeded out at this stage). Pre-trip training would also be required to prepare your body for the stress and strains of space travel.

Others may be knocked back for financial reasons, as was potential space tourist Lance Bass from the pop band 'N Sync. As the Soyuz rocket blasted off, in October 2002, Lance was left behind due to his failure to pay for the $28 million ticket.

You're kidding:
Space activities

Once space tourism progresses from brief flights to long stays in space, the range of possible activities is almost as endless as the universe.

You could enjoy the view – of Earth and out to space where the stars will appear brighter. You could perform acrobatics or simply enjoy moving around in zero gravity. Or try playing with objects such as water to see how they behave in space.

Once the crowds really start gathering in space, sports could be organised. Imagine the moves a gymnast or dancer could execute. Football stadiums would need to be huge to accommodate the long-distance kicks, and players could use each other's momentum to get to the ball first. Perhaps the Olympics of 2070 will be in space.

Entirely new sports could emerge that exploit the conditions of space and make Harry Potter's Quidditch look as exciting as a game of marbles.

Or you may make the choice not to be a space tourist. There are many who will always think a holiday is more about relaxation than adrenaline. After all, there aren't any nice beaches, interesting architecture or gourmet meals to be found up there. There is, however, plenty of space.

Adventure capitalists

So is this a growth area for investors? As well as companies such as Virgin Galactic and Space Adventures offering space flights, Lockheed Martin and NASA are working on a reusable space plane called the VentureStar,

HOW IT WORKS:
Space shuttlecock

SpaceShipOne doesn't blast off from a launch tower – it is carried by a 737-sized mother ship called *White Knight*, which can take off from any airport. This launch vehicle releases *SpaceShipOne* at the upper limits of current aircraft flights, where it glides for a few seconds before the pilot ignites the engines. In the thin upper atmosphere, the sleek spaceship can accelerate quickly to more than three times the speed of sound. Flying straight up, it reaches its farthest point from the Earth after a couple of minutes.

For its fallback through the atmosphere, *SpaceShipOne* reconfigures into a larger spacecraft by unfolding its wings and tails into a boomerang shape. Like a shuttlecock, air resistance slows its re-entry speed as it drifts through the thickening air. Once it reaches a height where the pilot can fold the spaceship's wings out again, it morphs into an aeroplane and flies like a glider back to the airport.

SpaceShipOne is made of Kevlar and composite materials. Rather than the highly-explosive fuel used by rockets that launch from the ground, it runs on a mixture of nitrous oxide (laughing gas) and fuel made of liquefied rubber that emits little pollution.

which could act as a tourist bus to space. Other companies, with names such as XCOR Aerospace, Rocketplane Ltd and American Astronautics, are also planning space flights, while Jeff Bezos, founder of Amazon.com, is planning a Texas-based rocket headquarters.

If you don't have the millions required to enter the space tourism race, maybe you could think about starting a series of space travel guides. Or, as a last resort, plan the first resort in space. Space hotels are likely to be the next innovation to follow the success of space flights. American hotel owner Robert Bigelow has an inflatable space station on the drawingboard, while the Space Island Group is at the design stage of a hotel that will orbit 700 kilometres above the Earth. One optimistic company is even selling real estate on the Moon. You can snap up an acre of lunar land for the bargain price of $60 from the Lunar Embassy company, but we don't like your chances of legally enforcing your ownership.

While increased space travel may sound like the potential for an environmental disaster – think of the massive energy needed – there may actually be environmental benefits. Gaia theory – that the Earth is a unique living organism – was born in the 1960s, when people first glimpsed pictures of a fragile blue planet in the vast darkness of space. Send enough people into space to see this view of Earth for themselves, and they might feel more inclined to look after it.

References

Internet

How stuff works, at <www.home.howstuffworks.com>.
New Scientist online edition, at <www.newscientist.com>.
Space Future, at <www.spacefuture.com>.
Virgin Galactic <www.virgingalactic.com>.
X-Prize, at <www.xprizefoundation.com>.

Text

Get Lost! 2005, no. 4.

The Next Big Thing is a Small Thing

Millions of molecules will affect millions of lives

It was kept out of the news, but in a remote laboratory in Nevada, a scientific experiment went horribly wrong. A team of scientists had used the new field of nanotechnology to produce microscopic robots. These micromachines, though brainless, could organise themselves in the way ants can work together to achieve swarm mentality. In the Nevada desert, the robots had organised themselves into flying black clouds, and they were deadly. The nanoswarm killed several scientists and threatened to escape to wreak havoc on an unprepared world.

This story is science fiction, part of the fantasy of Michael Crichton's 2002 novel, *Prey*. But the plot is based on public concerns that nanotechnology – the science of the very small – is developing at a rate faster than community understanding and government regulation can keep up with.

The fears of nanotechnology range from environmental to

Early days:
From church windows to buckyballs

Humans have formed objects since our distant ancestors shaped flint into axes. The quality of everything we make depends on the position of atoms – it's just that we've been less precise at arranging atoms in the past. The development of nanotechnology over the past thirty years has aimed to arrange atoms more cheaply, more efficiently and more friendly to the environment than past manufacturing methods.

Nanotechnology is new by name but not by nature. Chemists have been doing nanotechnology for centuries, reacting substances at the molecular level. Ancient artists used nanoparticles in the glazes on ceramic pots, while mediaeval craftsmen used nanoparticles of gold to colour stained-glass windows in churches. They had no idea about the science associated with this – they just knew that it worked.

Predictions that one day we would use tools to make smaller tools, which in turn could be used to make smaller tools, and so on down to the molecular level, were made by Richard Feynman in 1959. The term 'nanotechnology' was coined in 1974 by Nario Taniguchi to describe small machining tolerances. However, it was Eric Drexler, he of the grey goo, who in 1977 came up with the concept of using nanotechnology to manipulate molecules.

The 1981 invention of the scanning tunnelling microscope enabled scientists to finally see what they were doing. In 1991, Japanese researcher Sumio Iijima discovered needle-shaped molecules called carbon nanotubes. Six times lighter but 100 times stronger than steel, nanotubes are used as efficient transistors. Buckminsterfullerenes, or buckyballs, were discovered in 1985 and can be used in a

Early days cont.

variety of applications, including mopping up dangerous chemicals in the body, as tiny electrical components or even nanoscale ballbearings.

But just to revisit the nanofears once more, tiny airborne nanotubes could be inhaled and cause lung damage, while buckyballs have been shown to damage the brains of fish.

health concerns. Some technophobes see nanoparticles as the asbestos of the twenty-first century. Just as global warming is an unwanted environmental side effect of the industrial revolution, there could be unknown environmental side effects of the nanorevolution.

Prince Charles has said he is worried about the negative impact of nanotechnology. Even one of the pioneers of nanotechnology, Eric Drexler, warned of the problems that would result if self-replicating nanomachines were built to make and manipulate molecules. He wrote in his 1986 book, *Engines of Creation: The Coming Era of Nanotechnology*, that such tiny robots could spread from molecule to molecule, breaking down matter into formless grey goo. Asteroids bombarding the grey goo Earth as it floated lifelessly around the Sun would spread the mortal molecules through the universe until all matter was devoured. Keep in mind that when Eric described this he was writing science, not science fiction as Michael Crichton was.

Canadian bioethicist Peter Singer has different concerns. He says that while investment in and the number of scientific publications about nano-technology are skyrocketing, the number of papers and discussions covering its policy and ethics has flatlined. Rather than slow down the science, he says we need to speed up the social discussions to avoid a backlash reminiscent of that associated with genetically modified crops, which could rob us of the potential benefits of nanotechnology. And the potential of the tiny is far larger than you can imagine.

So do we run from nanotechnology or invest in it? We think the latter, and will conclude this book with a discussion of the next area of innovation likely to

HOW IT WORKS:
Not seeing is believing

Throughout history, the usual way to make an object has been to use a top-down approach. Manufacturers start with a lump of material and remove parts, grind it, melt it and generally shape it into something useful. Nanotechnology also uses a top-down approach, such as in the computer industry when tiny microprocessors are etched using microscopic tools. But it also uses a bottom-up approach, bringing together molecules to form objects in which the position of every atom is known. Another form of nanomanipulation is to build up layers of molecules, such as spray painting with atoms.

At the nanoscale, the world is weird. The boundaries between physics, chemistry and biology no longer exist. Quantum mechanics describes the behaviour of matter in totally different ways to what we're used to. If things in the real world behaved as they do at the atomic scale, you wouldn't be surprised to see cars bend around each other rather than colliding, or people walking through a solid brick wall.

While you can touch, smell, taste and see objects in the physical world, the atoms that make up objects are undetectable to your senses. Objects become invisible to the eye below about one-tenth of a millimetre, but nanoparticles are a few hundred times smaller than visible light – so forget about using a microscope to see what you're doing. Researchers see atoms using a technique that sweeps a very fine, charged needle over the material. Measuring the change in voltage allows an image to be built up at electron scale.

The next trick is to move the atoms. Opposable thumbs, long-nosed pliers or tweezers are all useless at this scale. Again, a charged needle nudges atoms into place, or picks them up in much the same way a giant magnet can move a car around a junkyard.

However, building anything one atom at a time would take about a billion years. This is where molecular self-assembly comes in – and the fear of Michael Crichton's *Prey*. But self-assembly is not about intelligent mini machines reproducing themselves. Rather, it is about a molecule knowing it should react with another. There's nothing scary about mixing molecules that snap together in a predesignated way – it's much the same process as reacting hydrogen and oxygen together: the result is water.

make millions, save millions and affect millions.

Small scale, big impact

You're likely to see – or rather, due to its size, not see – nanotechnology in many future applications.

Tiny valves in every cell of our body regulate internal water pressure. One of these so-called aquaporin molecules can filter a million molecules of pure water each second, removing sodium and chloride atoms that combine to form salt. Re-create these tiny, evolution-perfected water filters and you could have billions of them filtering seawater to produce pure drinking water.

Much smaller and lighter materials will enable the production of much smaller and faster computers. Storage capacity could increase a millionfold if information could be burnt onto disks as nanosized dots. And imagine computer readouts being displayed not on a large screen, but on a pair of glasses. That would be so cool, they'd need to be sunglasses.

Clothing made with metallic fibres with embedded electronics could sense changes in the weather and modulate airflow to keep you cool or warm. Nanomaterials in bricks could change the amount of air flowing into a building, depending on the temperature outside. If the building detects any damage, such as rust, it could release chemicals to repair cracks or corrosion.

Athletes will benefit from limb braces that sense when an injury is about to occur – and harden or relax accordingly. If an athlete is injured, future surgeons could just bang on a prosthetic limb that interacts with the body on a nanoscale.

Breaking matter into millions of particles increases its surface area without increasing its volume and so increases matter's ability to absorb, ignite or react – just as breaking up an ice cube exposes more of the ice to the air and makes it melt faster. Such properties could, for example, revolutionise the efficiency of drugs.

Intelligent drugs or medical devices could even act like antibodies, roaming the body to target bacteria or cancer cells.

Super-strong, light and durable materials would revolutionise transport, leading to more efficient aircraft and more environmentally friendly cars.

All of these futuristic visions are being discussed seriously by nanotechnologists. Seriously. Most far-fetched is the nanomaterial proposed for an elevator to space, which would be built of incredibly strong cables tethered to a satellite. Going up . . . way up.

Right under your nose

You won't have to wait long for nanoproducts – many are already here. The world's first nanomachine appeared in Australia in the 1990s. Researchers designed and built a device to detect small amounts of chemicals that could, for example, diagnose blood samples within minutes. The biosensor is just a few atoms wide, but is able to detect changes in concentration as tiny as the increase in sweetness that would result from adding a single sugar cube to Sydney Harbour.

Thick, white zinc cream was

not see-through, as is today's sunscreen, because of its large particles that scatter light. But in 1999, Terry Turney from CSIRO in Melbourne invented a modern, invisible version of zinc cream. The invisible cream contains a fine powder of zinc nanoparticles that block out harmful ultraviolet light but let through visible light so the cream appears transparent.

Similar radiation-absorbing properties of nanoparticles allowed Geoff Smith and colleagues at the University of Technology, Sydney, to develop the most affordable of a new breed of windows for buildings. They sandwiched between panes of glass tiny particles of a plastic called lanthanum hexaboride, which absorbs infrared heat without reducing the amount of visible light shining through. The result is cooler rooms that defy the need for airconditioning.

The University of Queensland physicist Michael Harvey was trying to develop low-cost antireflection coatings but accidentally discovered a way to stop glass from fogging up. His product, XeroCoat, applies an antifogging property to anything from mirrors to goggles to car windscreens. It works thanks to nanosized pores sucking up water that would otherwise form droplets on the glass surface.

If you have picked up a pre-mixed salad from the supermarket in the past ten years, you are likely to have held a plastic pack made of many thin polymer layers punctured by uncountable nanopores. Each tiny hole is permeable only to selected gases – for example, keeping in carbon dioxide but keeping out oxygen, which turns greens brown. Nanotechnology certainly is in its salad days.

Nanodollars

Spending on nanotechnology research and development in 2002 was $3 billion worldwide. In the 2005–06 financial year, nearly $12 billion will be spent worldwide; in Australia, nanotechnology spending will be $130 million, more than a quarter of it at CSIRO.

However, American venture capitalist investment in nan-otechnology halved between 2002 and 2004 – interests lie somewhere between wow products available now, such as stain-proof clothes, and the long-term promises of a nano-revolution. Nonetheless, total spending on nanotechnology by the private sector will, in 2005–06, exceed government spending for the first time in nanotechnology's short history.

The US National Science Foundation estimates that nanotechnology will be a trillion dollar industry by 2015. But nanoproducts, even though around today, may not cause a revolution for some years to come. Experts estimate that computers, batteries and medicine may widely benefit from nanotechnology by around 2012, energy production and transport will be influenced around 2025, while nanorobots won't appear until at least 2045.

References

Internet

Australian Academy of Science Nova, at <www.science.org.au/nova>.
CSIRO Nanotechnology Emerging Science Area, at
<www.csiro.au>.
Foresight Nanotech Institute, at <www.foresight.org>.
The Helix, online edition at <www.csiro.au/helix>.
Zyvex <www.zyvex.com>.

Text

Australian Anthill, Issue No. 9, April/May 2005.

Index